Modeling and Use of Context in Action

Modeling and Use of Context in Action

Edited by

Patrick Brézillon
Roy M. Turner

iSTE

WILEY

First published 2022 in Great Britain and the United States by ISTE Ltd and John Wiley & Sons, Inc.

ISTE Ltd
27-37 St George's Road
London SW19 4EU
UK

www.iste.co.uk

John Wiley & Sons, Inc.
111 River Street
Hoboken, NJ 07030
USA

www.wiley.com

Library of Congress Control Number: 2022934081

British Library Cataloguing-in-Publication Data
A CIP record for this book is available from the British Library
ISBN 978-1-78630-829-0

Contents

Preface

Context has always played an important, if little understood, role in human intelligence. This is especially true in human decision-making and communication. An individual's awareness of their own context as well as that of others with whom they interact allows many assumptions to be made about the discussion, the environment and/or the problem at hand. This allows many important aspects of human interaction to remain implicit when the communicants are in a common context, or alternatively, in different but mutually understood contexts. Otherwise, all assumptions would have to always be explicitly spelled out, which is a truly burdensome task for everyday communication.

Brézillon and Abu-Hakima (1995), in their report on the IJCAI-93 workshop "Using Knowledge in its Context", provide a summary of the discussions between the participants before (by electronic mail) and during the one-day workshop. Brézillon and Abu-Hakima clearly show that the notion of context was far from defined and is dependent in its interpretation on a cognitive science versus an engineering (or system building) point of view. The identification of these two viewpoints made it possible to go one step further than previous workshops. Once a distinction is made on the viewpoint, we can achieve a surprising consensus on the aspects of context, mainly the position, the elements, the representation and the use of context. However, despite the consensus on the aspects of context, agreement on the notion of context has not yet been achieved. Indeed, we still continue to see the two opposing views on context seen at the IJCAI-93 workshop. This book is an attempt to reconcile these two views.

The notion of context has been defined by a number of authors. Bazire and Brézillon (2005) presented a study on 166 definitions of context, but now we have 268 definitions. The results of their study show that context is specific to a given situation after answering the following questions: Who (i.e. the actor of the action)? What (i.e. the focus)? Where (i.e. the spatial location)? When (i.e. the temporal location)? Why (i.e. the intentions, the goals and even the emotions of the actor)? and How (i.e. the procedure needed to realize the action)? This study showed that the optimal order of presentation is action–object–agent–location, which gives a primary role to the action that confirms the necessity to consider the goal as a part of context. Webster's Dictionary defines context as "the whole situation, background or environment relevant to some happening or personality". This definition suggests that the context is always tacit and is rarely mentioned explicitly. Another definition of context, which is more operational in computing, is: "Context is what constrains a focus without intervening in it explicitly" (Brézillon and Pomerol 1999). Some researchers and practitioners have proposed singular and often narrow definitions, and focus on some aspects of context that can be identified from data, information or knowledge obtained through sensors that are specific to their domain or discipline (mainly around the context-aware computing community). In problem-solving, the context inherently contains much knowledge about the situation and environment of the problem. The context indeed constrains the focus, but conversely, the focus allows for the specification of the relevant contextual elements. For example, a dead battery in a car that has been parked overnight in freezing temperatures constrains the focus of a diagnostician, and it also specifies the contextual elements they must consider. Such a context has entirely different diagnostic implications than one where the car is in operation when the battery dies. Therefore, the effect of context on problem-solving and decision-making can likewise be very significant.

Context also introduces the expectation of a behavior that, by convention, goes along with a specific situation. Turner (1998) exemplifies this by alluding to the fact that when an individual enters a library, conversation is habitually reduced to a whisper. As the context changed (when the subject entered a library), new behaviors (whispers) were instantiated for use in the new context. Upon leaving a library, the individual's context changes again, and the prior behavior is no longer enforced.

Context affects virtually all aspects of behavior in animals, humans and computer systems. It affects how we understand the world, communicate with others, and plan and carry out our actions. It affects how computer systems should behave so that they act appropriately for their situation and their users.

We can find old references in some disciplines. For example, in mathematics, Frege (1892) identifies three different contexts: ordinary, direct and indirect contexts. Several theories also refer indirectly to context, such as activity theory (Leont'ev 1978), situation theory (Lave 1988) and distributed cognition (Flor and Hutchins 1991). It is generally acknowledged that John McCarthy brought the role of context in logic to prominence in 1993. The basic relation he introduced, ***ist(c,p)***, means the proposition ***p*** is true in the context ***c***. In other words, "context captures all that is not explicit in ***p*** but is required to make ***p*** a meaningful statement". There are several important conclusions in McCarthy's paper:

– a context is always relative to another context;

– contexts have an infinite dimension;

– contexts cannot be described completely;

– when several contexts occur in a discussion, there is a common (shared) context above all of them into which all terms and predicates can be lifted.

Note that in this formal approach, context is supposed to have a discrete nature, while in a cognitive approach there is a unique context of interest – the context of communication and interaction – which evolves continuously with the interactions.

In other disciplines, researchers consider it pointless to directly model context (one argument is the infinite dimension of context) and an external approach is preferred. For example, knowledge engineers introduced screening clauses in rule-based systems based on external control knowledge (not called context). For example, Clancey (1983) presents the following rule of MYCIN:

IF

1. The infection is meningitis,
2. Only circumstantial evidence is available for this case,
3. The type of meningitis is bacterial,
4. *The age of the patient is greater than 17 years old*, and
5. The patient is an alcoholic,

THEN

Prescribe tetracycline.

The fourth clause has nothing to do with the identification of the meningitis, but rather avoids firing the rule for children because tetracycline discolors growing teeth. Here, there is a unique context of interest that is intricately linked to the reasoning.

Contextual effects have been studied in many different disciplines over the years but usually without too much interaction between researchers in the different fields. This began to change in a small way in the mid-1990s with the 1993 *International Joint Conference on Artificial Intelligence* on modeling context in knowledge representation and reasoning, which sought to bring together some of the various threads of context-related research. The interdisciplinary focus on context as a subject of research truly began in 1997, when the *First International and Interdisciplinary Conference on Modeling and Using Context*, CONTEXT 1997, was held in Rio de Janeiro.

The two co-editors of this book were at the origin of the CONTEXT conference series. Our idea was to have a real interdisciplinary conference covering any discipline concerned with modeling and use of context. Disciplines ranged from philosophy to application development. However, over the last 25 years, the rapid pace of new technology resulted in the increasing importance of applications, and theoretical aspects of context without immediate spin-offs were of little interest to developers. As a result, for the last four years, the focus of the CONTEXT series has moved toward the pragmatic aspects of modeling and using context in order to serve all researchers and practitioners who develop concrete solutions based on any conceptual framework. Thus, we expect the audience of the book to be primarily those interested in developing pragmatic, operational frameworks for problem-solving by intelligent systems, not necessarily those interested in the more formal aspects of modeling context.

Conferences, after Rio de Janeiro (1997), have been held in Trento, Italy (1999), Dundee, UK (2001), Stanford, California, USA (2003), Paris, France (2005), Roskilde, Denmark (2007), Karlsruhe, Germany (2011), Annecy, France (2013), Larnaca, Cyprus (2015), Paris, France (2017) and Trento, Italy (2019), with the next conference planned for 2023. As the Covid-19 pandemic continued throughout 2021, a special issue of *Modeling and Using Context* was created in place of the conference. The lessons learned are that: (1) researchers come from different disciplines and wish to share ideas with researchers in other disciplines for solving similar problems by using context; and (2) the two opposite views on context, identified in 1993, are always present. The engineering viewpoint (the "hard science" following McCarthy and Giunchiglia), for whom contexts are discrete objects, and the cognitive viewpoint (the "soft science" following Simon and Newell), for whom there is a unique context evolving continuously.

Nevertheless, the focus of research on context has clearly evolved since 1997, in large part due to new technologies being developed and the resulting multiplication of "applications" communities. Within these new communities, researchers develop, at the core of their work, operational models of context for use in innovative real-world applications. As a result, CONTEXT conferences over the years have showcased papers from some of the leading thinkers on the subject, as well as newcomers from fields as diverse as artificial intelligence (AI), neuroscience, management and computer applications, and have hosted world-renowned speakers. We thus observed important changes in the themes of the CONTEXT conferences, such as "back to the root of context" (CONTEXT-15) concerning more particularly the fundamentals of context, and CONTEXT-17 focusing on "putting the context in the machine" to open the door to:

1) proposing a unified conceptual framework for all scientific disciplines for exchanging practical experience on modeling and managing context in real-world applications;

2) providing new communities (e.g. context-aware computing, mobile computing, ubiquitous computing, HCI (human–computer interaction), Internet of Things, etc.) a top-down approach to complement their bottom-up approach to modeling and managing context in real-world applications;

3) focusing on emerging domains, such as learning, e-health, mobility, human/social sciences (e.g. psychology), recognition, innovation, weak signals, context effects, knowledge management, case-based reasoning, user traces, collective/group work, quality of context, web services, Big Data, information and communication technology (ICT), and so on.

Among the lessons learned over the 11 CONTEXT conferences, we note that numerous disciplines are concerned with context but in very different domains. Beyond the heterogeneity of these approaches of context, a study of the keywords in the Proceedings from 1997 to 2017 allowed the identification of three general themes, namely communication, reasoning and activity but with varying importance over the 20 years (Brézillon and Turner, this book): respectively from (45%, 16%, 39%) in 1997 to (25%, 10%, 65%) in 2017.

Consequently, the notion of interdisciplinarity, which is the basis of our context community, now has its center of gravity more toward applications. New communities have arisen centered around getting applications to be more aware of their environment and their users through an explicit representation of their contexts. Thus, there is a natural shift in the interdisciplinary focus of our CONTEXT conferences, opening the door to new communities to share with them our interdisciplinary view on context by providing a unifying, interdisciplinary forum and framework for studying and applying context for all those interested in application areas prefixed by "context-": context-aware, context-based, context-sensitive, etc.

There are several reasons for this shift in emphasis over time. A key reason is the rapidly growing importance of ICT, which uses preferentially bottom-up rather than top-down approaches. Indeed, ICT relies on a notion of innovation in three steps: (1) a creation or invention, (2) the need for immediate implementation and (3) appropriation of the innovation by end-users. Such a notion, like the Jugaad

Innovation (Radjou et al. 2013), is the opposite of the approach that uses logic as the unique modeling tool of human intelligence.

To take into account this new reorganization of the science landscape, we propose consideration of – instead of the three initial themes – the three following revised approaches of modeling and using context: formal, cognitive and pragmatic. Thus, our community of context should reorient its goals to focus on:

– providing all scientific disciplines with a platform and a forum for exchanging practical experience modeling and managing context in real-world applications;

– providing new communities (e.g. context-aware computing, mobile computing, ubiquitous computing, HCI, the Internet of Things, etc.) with top-down ideas and models of context to complement their bottom-up approaches for modeling and managing context in real-world applications;

– informing, guiding and providing a forum for emerging areas and areas just beginning to seriously consider context, such as learning, e-health, mobility, human/social sciences, knowledge management, quality of context, web services, etc.;

– providing insight into theoretical aspects of context by learning and generalizing from studying and generalizing how context is modeled and used in real-world systems.

This book aims to account for this move by the interdisciplinary community that is interested in modeling and using context in real-world applications. It is an important challenge for such a community, since researchers have similar problems at a strategy level but only differences in their domains at the operational level.

With the move being made by the context community across disciplines, a special emphasis on context and AI is needed. Researchers in AI have used context for a long time, for example de Kleer (1987) in ATMS, McDermott (1982) in X1/RCON, Laird et al. (1987) in SOAR, Hendrix (1975) for the partition of semantic networks, Guha (1991) in the CYC project. Winograd and Flores (1986) considered that the ability to act comes from our familiarity with the tool (mental representation), not from our knowledge of the tool, i.e. how the tool was used in context. Indeed, the notion of mental representation of operators is an important element (for user modeling, task realization, decision-making, etc.). The relevant space of choice is rarely the space of all logical possibilities but rather of an orientation within a context. Thus, a situation is established by an actor and meaning always derives from a contextual interpretation rooted in this situation. The actor's mental model will be correct in some ways and incorrect in others. Context plays a role in a space of possibilities that allows us to attend to both explicit information and tacit knowledge. Thus, context is composed of projected perspectives on the world, and the set of known alternatives already explored.

Engel et al. (2021) identify the following set of characteristics of AI: (1) *experimental character*, which refers to AI outcomes being non-deterministic but probabilistic; (2) *context sensitivity*, which refers to AI solutions being only as good as the data their context provides; (3) *black box character*, which refers to AI systems, especially in the field of deep learning, facing challenges in delivering explanations to humans on what happens between data input and AI output; and (4) *learning requirements*, which refers to AI solutions constituting entities that, just like humans, need to learn and develop experience to eventually improve their performance over time.

"Big Data" and other data sources for AI development have led to the development of formal tools for AI systems to develop methods of data acquisition for deep learning and machine learning to begin to directly engage with real-world applications. AI systems could very soon have reasonably natural conversations in certain domains. However, to do this, they need to be grounded in context (see Chapter 5).

Context involves many interconnected layers of accumulated knowledge (see the onion metaphor in Chapter 2) that humans acquire and apply in conversation with little effort. However, computers cannot yet accumulate such knowledge without an efficient modeling of context. This contextual knowledge includes the purpose of a conversation, information on its interlocutor and past experience with them, the lessons learned from previous interactions with other interlocutors with the same purpose, general information about the world as it relates to this purpose, and so much more.

Contextual awareness describes the ability of AI systems and IT components to respond to requests based on their knowledge of the situation. The more contextual information an AI system gathers, the more effective it becomes. Everything from wearable devices to Internet-connected machines such as refrigerators or virtual assistants will become more connected to contextual information. Indeed, emerging communities (e.g. context-aware and ubiquitous computing, etc.) aim to get applications to be more aware of their environment (i.e. part of the context) and of their users through explicit representation of their contexts.

Making context explicit in such approaches is supposed to bring a solution to the representation (and use) of a dynamic organization of knowledge according to the focus, a way to speed up reasoning and optimize functioning in a changing environment, as well as activity modeling and autonomous recognition and behavior. This supposes that context is interrelated with other knowledge and reasoning, just as other knowledge and reasoning are themselves interrelated. Context is anchored in the real world and cannot be expressed in simplified models of reality.

Patrick BRÉZILLON
Sorbonne University, France
Roy M. TURNER
University of Maine, USA
March 2022

References

Bazire, M. and Brézillon, P. (2005). Understanding context before using it. In *Modeling and Using Context (CONTEXT-05)*, Dey, A., Kokinov, B., Leake, D., Turner, R. (eds). Springer Verlag, Berlin.

Brézillon, P. and Abu-Hakima, S. (1995). Using knowledge in its context: Report on the IJCAI-93 Workshop. *AI Magazine*, 16(1), 87–91.

Brézillon, P. and Pomerol, J.-C. (1999). Contextual knowledge sharing and cooperation in intelligent assistant systems. *Le Travail Humain*, 62(3), 223–246.

Clancey, W.J. (1983). The epistemology of a rule-based expert system – A framework for explanation. *Artificial Intelligence*, 20(3), 215–251.

Engel, C., Ebel, P., van Giffen, B. (2021). Empirically exploring the cause-effect relationships of AI characteristics, project management challenges, and organizational change. *Proceedings of the 16th International Conference on Wirtschaftsinformatik*, Essen, Germany.

Flor, N. and Hutchins, E. (1991). Analysing distributed cognition in software teams: A case study of team programming during perfective software maintenance. In *Proceedings of the Fourth Annual Workshop on Empirical Studies of Programmers*, Koenemann-Belliveau, J., Moher, T.G., Robertson, S.P. (eds). Ablex Publishing, Norwood.

Frege, G. (1892). Sinn und Bedeutung. Zeitschrift fuer Philosophie und philosophische Kritik, 100: 25–50. In *Philosophical Writings of Gottlob Frege*, 3rd edition, Geach, P. and Black, M. (eds). Blackwell, Oxford.

Guha, R.V. (1991). Contexts: A formalization and some applications. MCC technical report, ACT- CYC-423-91, December.

Hendrix, G. (1975). Expanding the utility of semantic networks through partitioning. *Proceedings of the Fourth IJCAI*, 115–121, Tbilisi, Georgia.

de Kleer, J. (1987). An assumption-based truth maintenance system. In *Readings In Nonmonotonic Reasoning*, Ginsberg, M. (ed.). Morgan Kaufmann, Los Altos.

Laird, J.E., Newell, A., Rosenbloom, P.S. (1987). Soar: An architecture for general intelligence. *Artificial Intelligence Journal*, 33, 1–64.

Lave, J. (1988). *Cognition in Practice: Mind, Mathematics, and Culture in Everyday Life*. Cambridge University Press, Cambridge.

Leont'ev, A.N. (1978). *Activity, Consciousness, and Personality*. Prentice-Hall, Hoboken.

McCarthy, J. (1993). Notes on formalizing context. *Proceedings of the 13th IJCAI*, 555–560.

McDermott, J. (1982). R1: A rule based configurer of computer systems. *Artificial Intelligence Journal*, 19, 39–88.

Radjou, N., Prabhu, J., Ahuja, S. (2013). *The Innovation Jugaad*. Éditions Diateino, Paris.

Turner, R.M. (1998). Context-mediated behavior for intelligent agents. *International Journal of Human–Computer Studies*, 48(3), 307–330.

Winograd, T. and Flores, F. (1986). *Understanding Computers and Cognition*. Ablex, Norwood.

Introduction

This book has its origin in the CONTEXT series of biennial interdisciplinary and international conferences on modeling and using context, which began in 1997. Over the past 25 years or so, there has been a clear evolution in the interests of the community for modeling and using context, from formal aspects (e.g. natural language processing and logic) to pragmatic approaches, in large part due to the dynamic development of information and communication technology over that time. The consequence is a change from top-down to bottom-up research where "real" problems need rapid pragmatic solutions that are validated and justified by the results, even without a well-established theory.

This book aims to provide evidence for the success of pragmatic approaches, which in any case are different but still related to formal approaches. We have selected 10 relevant contributions across a very large spectrum of domains, from technological domains (e.g. subways) to more fuzzy domains (e.g. emotions). The first three chapters lay the foundations for pragmatic approaches, and the following chapters for applications.

Chapter 1 presents the challenges of pragmatic research on context modeling and use, mainly in artificial intelligence (AI) for problem-solving by agents (human and AI), and discusses existing and past bottom-up approaches. The agent viewpoint makes it possible to bring together context data from the external world, information (for agents and groups of agents) and knowledge (inside agents), and modeling reasoning (human and AI). Three examples of context-based research addressing applications in the real world are presented.

Introduction written by Patrick BRÉZILLON and Roy M. TURNER.

Chapter 2 presents the results of 25 years of research on modeling and using context in a number of real-world applications. The 12 lessons learned throughout these years of research show that it is possible to model context, develop a software (called contextual graphs) and use the software for modeling operational knowledge in applications in which the process (task realization, decision-making, problem-solving) is at least as important as the result. The success comes from a twofold approach: at a conceptual level and an implementation level. The software provides such a general solution that the applications concerned cover a large spectrum from technical domains (e.g. power systems and subways) to more fuzzy domains (e.g. winemaking and medicine).

Chapter 3 develops an approach toward pragmatic context-based intelligent systems, developing the ideas presented in Chapter 1. Context-based intelligent systems focus on problems and the context in which the problems are to be solved. The authors argue that the pragmatic approach is particularly useful for building systems to solve real-world problems. The requirements needed for pragmatic context-based systems are presented in general as well as in particular for two broad classes: context-based intelligent assistant systems (CIASs) and context-based intelligent autonomous agents (CIAAs). They use their own work in pragmatic context-based intelligent systems, for example.

Chapter 4 presents the lessons learned across 10 years of research on context effect-based education with the implementation of the TEEC (Educational Technologies for Teaching in Context) project. This project conceptualizes a techno-pedagogical model where context occupies a central role. The key principle of the approach is the involvement of student groups living in Quebec and Guadeloupe, sharing the same language, and collaborating on subjects that have local specificities. Collaboration is done synchronously and asynchronously through existing digital platforms. Several experiments have been completed successfully. Among the digital tool developed, there is a context-aware intelligent tutoring system.

Chapter 5 concerns a pragmatic approach to context-based intelligent reasoning, context-mediated behavior (CMB). CMB explicitly represents contextual information as *contextual schemas* (c-schemas), which represent classes of situations that have implications for how an agent should behave. Contextual knowledge contained in c-schemas includes descriptive knowledge about the situations represented and prescriptive behavior about how to behave. An agent's current situation is diagnosed to be one or more c-schemas, whose knowledge is merged to guide the agent while in the context.

Chapter 6 describes how the use of context can help identify the emotion being expressed by a person when they are vocalizing sounds without understandable words. The objective of this research is to determine whether the use of context can help in making such inferences. The technique used comes from the field of sentiment analysis to model the context of the actor vocalizing the sounds. Mapping of keywords selected from a textual description of the actor's context to emotions, as contained in a sentiment analysis library, can improve the selection accuracy by approximately 20% over the published state of the art for 17 possible emotions. This chapter describes the context modeling process used and the overall system designed to perform the affect recognition.

Chapter 7 presents a proactive approach to contextual reasoning, context-driven behavior, a descendant of the research presented in Chapter 5. Existing frameworks and models of contextual reasoning are primarily reactive in nature, i.e. without consideration for future contexts. However, when changes occur, the agent must identify new contexts and adapt its behavior accordingly. The novel approach proposed allows an agent to use its contextual knowledge proactively. That is, as opposed to reacting to contextual changes, this approach allows an agent to make predictions about future contexts it expects to encounter, and then use knowledge of those contexts to drive aspects of its behavior before it enters into them.

Chapter 8 discusses a context-based approach to data discovery for anonymization in a digital world where the amount of data is increasing exponentially. The goal of data discovery is to inspect metadata and real data in different databases to discover personal data and provide comprehensive listings of sensitive fields storing that data. The authors discuss how to consider context in personal data discovery to improve the anonymization process. The use of the contextual graphs formalism allows procedures to be improved for sensitive personal data discovery used in anonymization processes.

Chapter 9 presents a dynamic model of situated management learning, a tool for teaching, practicing and researching management. The model leverages three different and complementary ways of understanding, modeling and using context grounded, respectively, in practice, values and theoretical insights. The examples presented use a contextual graph to represent practical knowledge, a values table to represent political, moral and ethical knowledge, and an insights table to represent theoretical knowledge. The model has been used in teaching accounting at a business school in Paris and the results are discussed.

1

Pragmatic Research on Context Modeling and Use

Pragmatic approaches to solving problems in artificial intelligence arise due to rapid technological evolution as well as the complexity of real-world tasks. These operational, bottom-up approaches are based on representing how to behave in specific contexts. In this chapter, pragmatic approaches are discussed from the perspective of agents (human and AI) using data, information and knowledge in context. Information is what is interpreted by an agent from data about the world, and what is exchanged with other agents. Information is highly context-dependent and under the control of the knowledge that is in the agent's "mind". Context plays a pivotal role in the development of information and knowledge. Setting data, information and knowledge in context allows new approaches to modeling (human and AI) reasoning. Based on its contextual knowledge (the part of its knowledge concerning the current focus), a reasoner interprets and uses information and produces new information and knowledge. Concretely, the pragmatic approach is shown in three examples of context-based research addressing applications in the real world. Operational knowledge in these approaches leads to contextualization of reasoning and opens the door to context-based AI systems.

1.1. Introduction

As artificial intelligence (AI) has matured from its early roots in research focused on "weak" or general-purpose methods, the field has become increasingly focused on "strong" (knowledge-based), pragmatic approaches to solving problems (see Chapter 3). While research on formal, theoretical topics is still very important, it has become apparent that in order to solve complex real-world problems, a "top-down" approach from theory to system is impractical.

Chapter written by Patrick BRÉZILLON and Roy M. TURNER.

AI research has, over time, also become increasingly aware of the role context and contextual knowledge plays in problem-solving. This has evolved from tentative approaches such as the situation calculus (McCarthy and Hayes 1969) and dividing rules in expert systems and knowledge-based systems into context-specific packets (e.g. Chandrasekaran et al. 1979; Brézillon 1989) to formal approaches focused on context per se (e.g. Giunchiglia 1993; McCarthy 1993) to an entire community focused on modeling and using context, beginning with a workshop at the 1993 *International Joint Conference on Artificial Intelligence* (Brézillon and Abu-Hakima 1995). This community has grown since them, centered around the CONTEXT (*International and Interdisciplinary Conference on Modeling and Using Context*) biennial conference series. As with AI in general, this community has become increasingly focused on pragmatic approaches to solve real-world problems.

In this chapter, we discuss the pragmatic approach to context modeling and use in intelligent systems, and then examine three approaches we consider to be examples of pragmatic context-based systems and the lessons that can be learned from them for such systems in general. Chapter 3 takes a closer look at the requirements for pragmatic context-based systems going forward.

1.2. Pragmatic research on context

While natural language researchers have been concerned with context in a narrow sense for many years, context first appeared as an important challenge in AI itself at the *International Joint Conference on Artificial Intelligence* in 1993 (IJCAI-93). In his seminal paper, McCarthy (1993) introduced the relation *ist(c,p)*, which states that a proposition *p* is true in context *c*, and concluded that: a context is always relative to another context; contexts have an infinite dimension; contexts cannot be described completely; and when several contexts occur in a discussion, there is a common context above all of them into which all terms and predicates can be *lifted*. As a result, we cannot speak of context outside of its own context.

Context is a concern not only for AI theorists and practitioners but also for a wide range of disciplines, including psychology and cognitive science, linguistics and natural language processing, neuroscience, anthropology, sociology, organizational behavior and philosophy. This community of researchers was fostered and made concrete by the CONTEXT series of biennial conferences, beginning in 1997 and continuing to the present day. Other specialized research communities interested in context have arisen over the years (e.g. context-aware computing, ubiquitous computing, etc.), but CONTEXT remains the backbone of the interdisciplinary context community.

However, even though it is critical to the success of the interdisciplinary CONTEXT community, there is still no universal definition of context, as pointed out by Bazire and Brézillon (2005), who collected 166 definitions of context in the literature; as of 2021, the number had grown to 268. Similarly, there is no single definition of contextual knowledge, that is, the knowledge an agent needs about a context and about how to behave while in it. There is also the problem of determining which knowledge is contextual (related to the task focus) and which is not (i.e. external to the task focus). And this is not a static determination, since a piece of external knowledge may become contextual or vice versa as the situation changes.

There has been a shift in emphasis and concerns over time. At the first CONTEXT conference in 1997, roughly 40% of the papers related to communication (including NLP and most cognitive science approaches), 16% to reasoning (formal aspects) and 44% to what might be called "activity" (applications, user interfaces, etc.). Twenty years later, in 2017, this had changed to 25% being related to communication, 10% to reasoning and 65% to "activity". This was driven in large part by the rapidly growing importance of information and communication technology (ICT) over this time. Since ICT relies on innovations, immediate implementations and adoption by end-users, it is not concerned with formal aspects of reasoning or cognitive plausibility.

If we focus on a *pragmatic* view of context – how context and contextual knowledge should be modeled in and affect real-world applications and agents – then the definitions at least begin to converge. Here, we can view the agent's knowledge as its *mental model* of how to realize its task in the local environment, a model that is developed by accumulated experiences, that is, contextualized knowledge or operational knowledge. Such a mental model assembles, organizes and structures all the contextual elements (CE) encountered by the actor during their past experience with this task realization, decision-making or problem-solving. This is different from the traditional view of a knowledge base or axiom set divorced from how that knowledge is used. A mental model implies a focus on knowledge's operational and dynamical organization for the current context, changing as the context changes. It is different from agent to agent, and it is built from the agent's experience performing tasks in different contexts.

1.3. Role of context in AI systems

1.3.1. *Data, information and knowledge*

We can see the role of context in AI systems perhaps best from the *agent-based* perspective (Russell and Norvig 2020), to which much of modern AI subscribes. An *agent* is an entity that repeatedly accepts a percept from the environment, makes a decision and then takes action (see Figure 1.1). A *percept* can be sensor data, a message from other agents, communication from a user, etc., or it can be null (so the agent can be proactive as well as reactive). Except for the most rudimentary reflex agents, the decision-making process involves updating the agent's own internal knowledge based on the new percept, and the agent's decisions are based on this updated internal knowledge. Humans, animals and virtually all AI systems can be usefully conceptualized as agents.

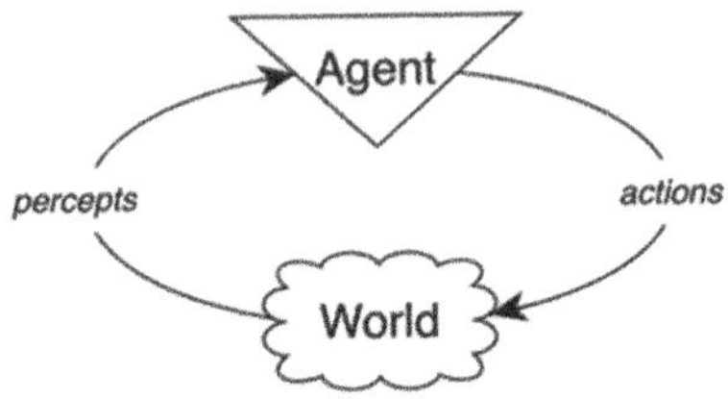

Figure 1.1. *An agent and the world*

What counts as percepts, decisions and actions will vary from agent to agent. For example, a typical deep-learning vision system has an image as its percept; its "decision process" is just the extremely complex function it embodies; and its action is the result of applying that function to the input image to classify it. An intelligent autonomous underwater vehicle (AUV) derives its percepts from its sensors; its decision process is one or more AI techniques, such as automated planning, behavior-based control, etc.; and its output is via its effectors and their effects on the world. An intelligent assistant system receives inputs from the world and its user; its decision process is comprised of one or more AI techniques; and its outputs are its recommendations, warnings, etc., to the user as well as any direct control it has over the world.

We can think of the information used by an agent as being of three types. The first and simplest kind is *data*: information that is agent- and context-independent, that is, objective. Examples of this are raw or filtered sensor signals (e.g. a voltage from a temperature sensor, a pixel's RGB values, etc.) and even some symbolic information ("temperature is 24°C").

When data are interpreted, processed, organized, structured or presented in a given context, it becomes *information* that is meaningful to the agent who receives it. Information in this sense is data that has been converted into a more useful or intelligible form for helping humans or AI agents in their decision-making process. Context-dependent information has meaning that may vary with the context or the individual reasoning about it. For example, "it is warm" is an assessment based on data (the raw temperature, the feeling of the air on our skin, etc.), but relative to a particular context and agent; it could refer to 16 °C in Paris in March or −12°C in January in Maine.

The third kind of information is *knowledge*. We consider this to be information that is both context-dependent *and* integrated into the rest of the body of knowledge possessed by an agent, often as a result of reasoning about it in relation to the rest of the knowledge. Knowledge is information containing wisdom, so to speak. For example, "386" is data, "your marks are 386" is information and "You have done well because of your hard work" is knowledge. It is knowledge upon which an agent bases its decisions.

As shown in Figure 1.2, the three kinds exist in a hierarchy of sorts, with complexity, agent-centricity and integration with other information increasing as we move up the hierarchy. Data, which can come from either sensors or communication (from another agent or the user), is transformed into information by a process of interpretation. Information, which can also come from communication, is in turn transformed into knowledge by reasoning and by integrating it into the agent's other knowledge. As is also shown, information can come from interpreting other information, and new knowledge can also be produced by reasoning (i.e. by inferences) about existing knowledge.

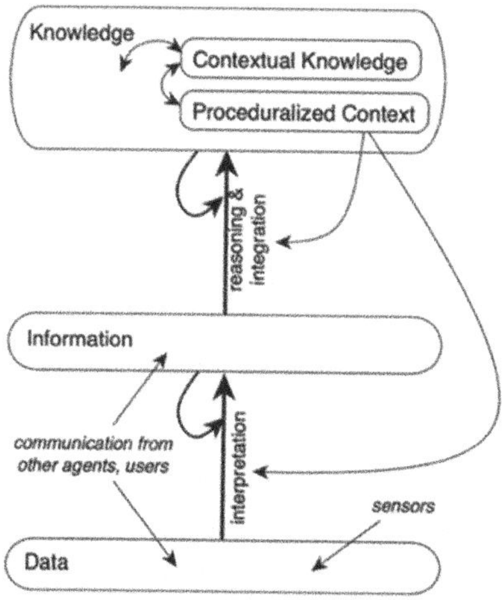

Figure 1.2. *Information hierarchy*

The problem with managing knowledge is that it is largely contextual with a strong experiential bias, which makes identifying, acquiring and replicating it difficult. Knowledge is of two interdependent types.

– Fact-based or information-based: this comes from fundamental science, derived from experiments, rules and principles commonly agreed upon by experts. This type of knowledge is found in models and software.

– Heuristic: this is knowledge of good practice, experience and good judgment. It is the knowledge underlying "expertise", rules of thumb or hypotheses about what usually works to achieve desired results, but it does not guarantee them. This type of knowledge pertains to humans and AI agents (see Chapter 3).

Many kinds of software use the bottom tiers in the hierarchy. For example, signal processing, data mining and databases, and some kinds of deep learning all take as input data or information and produce other data or information, usually for use and interpretation by a human. However, only humans and intelligent systems can be said to use knowledge in the sense of actually understanding the knowledge itself and integrating it into their own model of the world and ongoing processing.

1.3.2. *Contextual knowledge*

Context impacts the activities of all humans and real-world AI systems and, consequently, contextual knowledge must be used by them to function properly within their contexts. As shown in Figure 1.2, contextual knowledge is only one type of knowledge an agent has, and whether or not a piece of knowledge is contextual can change over time as the problem-solving focus changes. Knowledge not relevant to the current context has been referred to as *external knowledge* (see Chapter 2). Of contextual knowledge, there are two types: knowledge that is relevant to the current context, but that has not (yet) been instantiated with the specifics of the situation because it is not directly being used (but may have an indirect influence); and *proceduralized context* (Chapter 2), that is, the part of contextual knowledge that has been instantiated based on the task and particular situation at hand and that to some extent directs the agent's processing[1]. A piece of knowledge may move between proceduralized context and non-proceduralized context as the focus of problem-solving and/or the situation changes.

Intelligent agents are influenced by contextual knowledge in several ways. It can help a reasoner interpret information it has, producing new information. Simple examples of this would be using context-specific rules to infer from the fact that "depth is 10 m" and from the context of an agent's location (and corresponding data from undersea depth maps) that it is currently at an altitude of 30 m from the sea floor, or to infer from a machine being turned on and the context that the user is present that the user turned the machine on and, thus, meant for it to be on. Context

1 This reflects work by one of us (Chapter 2); work by the other (Chapter 5) does not use the term PC, but there are similarities in how contextual knowledge is used in that work.

can also help an agent interpret information to produce subjective knowledge by suggesting how to integrate it into the rest of the agent's knowledge.

For both humans and artificial agents, context limits the possibilities for interpretation, reasoning, decision-making and acting[2]. For incident solving on a subway line, the operator looks first at the contextual knowledge (status of the entire line, rush hour, how things are done on the other subway lines in correspondence with their line, etc.) in order to select the best contextual knowledge and means to the ultimate goal of ensuring the security of travelers. For a given agent, a change in mood implies a change of context, therefore possibly a different decision.

Context is relevant to Simon's (1960) key contributions to understanding the decision-making process. According to Simon and his later work with Newell (Newell and Simon 1972), decision-making is a process with three distinct phases.

– **Intelligence phase:** the decision-maker identifies/detects the problem or opportunity. A problem is anything that is not according to the plan, rule or standard, while an opportunity is any promising circumstance that might lead to better results. Information is gathered about a problem, formatted into a useful structure, then factored into the information about the context in which the problem has occurred. The context has a major role in decision-making, and information is required both about the problem and about the context in which the problem occurred. Without information about a problem or opportunity, the decision-making process cannot even start. Without information about the context in which the problem has occurred, no correct decision can be made. Information about the context is integral to the complete definition of the problem.

– **Design phase:** the design phase involves the search for alternative solutions to solve a problem. Information is a key ingredient in the generation of alternatives for decision-making. Each alternative solution is evaluated after gathering data about it. The evaluation is done on the basis of criteria to identify the positive and negative aspects of each solution. Alternative solutions may be difficult to identify in the case of insufficient information and may require a return to the intelligence phase.

– **Choice phase:** in this stage, alternative solutions are compared to find the most suitable one. This is not as easy as it seems because each solution presents a scenario and the problem itself may have multiple objectives. The lack of a satisfactory solution requires a return to the design phase. Based on the information about the suitability of the alternatives, a choice is made to select the best alternative.

---

2 This is true for larger groups of people, too; for example, a *paradigm* in Kuhn's (1970) view of scientific revolutions strictly limits the perception and reasoning of large swaths of the scientific community.

Each of these phases of decision-making requires knowledge about the context in which it is occurring. Problems or opportunities can only be understood in context, and the information needed to generate alternatives during the design phase comes in large part from knowledge about the context for the solution. Evaluation and choice of alternative is context-dependent as well, since an alternative that is perfectly fine in one context (going into the kitchen to get a glass of water at home) may be completely inappropriate in another (when in a restaurant).

1.4. Three examples of pragmatic research on context

1.4.1. *Introduction*

In this section, we compare three examples of pragmatic research in context-sensitive reasoning: context-based reasoning (CxBR), contextual graphs (CxG) and context-mediated behavior (CMB). These three context-driven approaches were developed in different contexts for addressing initially different objectives. The discussion here comparing them is an extension of work by Gonzalez and Brézillon (2008), who searched for a unifying vision of CxBR and CxG, or at least identification of either their overlap or the specificity of their domains of application. Here, we include CMB in the evaluation to support generalizing conclusions about properties and abilities needed for any pragmatic context-based reasoning approaches. CMB, as is also true of CxBR, is concerned with AI agents, while CxG systems interact with a human or simulated agent (see Chapter 3).

The initial comparison was based on 10 different criteria, with some indication of which one excels at each particular facet of performance. It focused the comparison on how each would represent human or agent tactical behavior, either in a simulation or in the real world.

Conceptually, these three context-driven approaches are not at the same decision level. This could provide an opportunity in the future to combine them synergistically for a more global approach to context-based reasoners. The inclusion of CMB here broadens the range of application of the systems studied, since CxBR specifically concerned human behavior in a tactical situation, while CMB was developed for AI agents acting autonomously or in multiagent systems at the tactical and strategic levels; and CxG, on the one hand, is at the tactical level with an experience base containing all the practices developed for realizing a given task, and, on the other hand, at an operational level for developing the specific practice of the contextual graph corresponding to the working context at hand. Apart from that, the three context-based approaches focus on task realization, decision-making and problem-solving. Context serves to help prune the search for solutions and/or to influence the solution selected or the decision made.

These context-based approaches aim to attack the most challenging problems in modeling human and artificial agent behavior through:

1) identifying a context and its relevant features. The classical frame problem is closely related to this issue;

2) deciding what is the most appropriate behavior (actions, practices) for the current context;

3) knowing when a change in context has taken place. Humans know this instinctively. However, it is difficult to generalize rules that dictate when to switch to another context as a result of environmental and/or internal events.

While these approaches share similarities, they are also quite different in how they model contexts and use contextual knowledge. Therefore, they address the above problems from different perspectives.

1.4.2. *Contextual graphs (CxGs)*

A *contextual graph* (Brézillon et al. 2000) is a directed, acyclic, series–parallel graph in which a path represents the realization of a task where the actions are undertaken in accordance with the current context. Actions aim to achieve a goal, while *contextual nodes* describe the possible contextual issues of a given focus (e.g. realization of a task) in different situations. The contextual graph contains the CE more or less related to the focus, and, among them, the *proceduralized context*, that is, the ordered set of instantiated CE corresponding to the given practice development at hand. The proceduralized context are the specific context *chunks* ready to be used in the actions selected for the focus. The proceduralization of the contextual knowledge along the path makes the links explicit, especially the causal and consequential links, between contextual knowledge chunks and, as such, the links become part of the proceduralized context. Thus, the proceduralized context appears as a kind of compiled (i.e. contextualized) knowledge that the system can easily “decompile” to explain its reasoning. Consequently, we can regard a contextual graph as representing the proceduralized context for any path (i.e. practices developed for the focus in different contexts), because for each context represented by a sequence of instantiated CE, the implicit reasoning about causes and consequences implies that the action to undertake is defined without ambiguity. In other words, each sequence of contextual nodes along a branch triggers some actions as a result of rationales that are not represented on the graph but are generally known and compiled in the user’s mind (contextualized knowledge). Thus, the CxG explicitly represents the reasoning jointly involved with its proceduralized context.

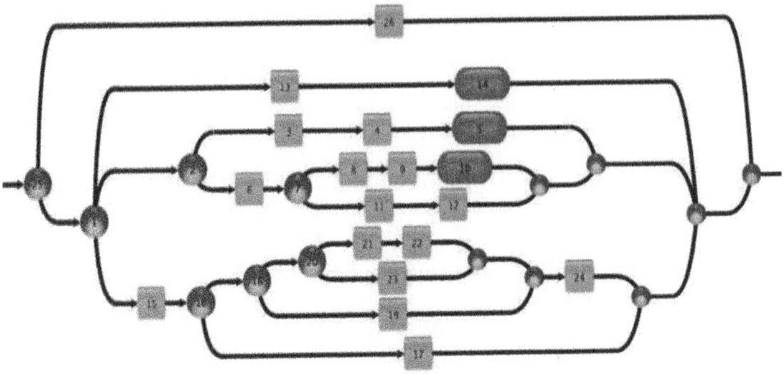

Figure 1.3. *Contextual graph organization. For a color version of this figure, see www.iste.co.uk/brezillon/context.zip*

Figure 1.4 depicts a contextual graph and its organization of CE. A CE is a pair consisting of a *contextual node* and a *recombination node*. In the figure, the circular nodes with a number are the contextual nodes, while the square ones are the action nodes and the elongated ovals represent the activities (independent contextual subgraphs). The contextual nodes require a decision or an information input, generally given by the user (but a system may also retrieve the required information), while the action or activity nodes perform an action or describe an action to be performed by the agent. It may also modify the environment and, by extension, the context. The smaller unnumbered circular nodes are the recombination nodes where a particular CE ends. Each recombination node is associated with a contextual node, and both correspond to a CE.

The purpose of contextual graphs is to represent, along each branch of the graph, an increasingly refined state of the problem that users try to identify in order to better target a solution. For example, operators in control rooms for subways want to work with the certainty of having entered the proper branch (i.e. the correct practice to develop), and they generally postpone actions until after the collection of CE. The proceduralized context evolves continuously with the practice development. Some CE enter while others leave the proceduralized context throughout the process of reasoning.

Recall that the organization of a contextual graph makes explicit the context and its dynamics for decision-making. This representation is more compact than decision trees and was well received by the operators in a subway line management application (Pasquier et al. 2000). The representation in a contextual graph of all the known ways for task realization or decision-making as practices allows a contextual graph to behave as an experience base, a basic attribute of context-based humans and AI agents.

Although the CxG formalism was used to model mainly experts' reasoning in task realization, decision-making and problem-solving, the collection of the instantiation of the CE, the generation of the proceduralized context and the execution of the actions and activities in the focus may be automated in an AI agent (as implemented in a CxG-based simulator).

1.4.3. *Context-based reasoning (CxBR)*

Context-based reasoning (Gonzalez et al. 2008) is based on three ideas.

1) A *situation* corresponds to a set of actions, procedures and expectations that properly address the situation. These sets of actions, procedures and expectations are called *contexts*.

2) Evolution of the situation requires a new set of actions, procedures and expectations to manage the newly emerging situation. Therefore, a *transition* to a new context must be done successfully.

3) A future situation is identified supposing that things in the current situation are limited by the current situation itself. This facilitates *situational awareness*.

CxBR encapsulates into contexts knowledge about appropriate actions, procedures and expectations as well as knowledge about possible new situations. CxBR is based on concepts such as: not all inputs are relevant; a situation may evolve toward a finite number of subsequent situations, decisions are made in contexts with a defined set of tools; and context changes depend on external events.

The hierarchical organization of CxBR is shown in Figure 1.4. CxBR contexts are organized as a graph, with each context possibly having multiple parents. The hierarchy consists of three levels: the mission context, the major contexts and the minor contexts (see Figure 1.4). There is only one mission context, and it serves to define rather than control the agent. Major contexts are the main control element, and while there may be several assigned to a particular mission, one and only one actively controls the agent at any one time. This is referred to as the *active context*. The minor contexts play a supporting role, and there can be several levels of minor contexts, starting with the sub-contexts.

A full description of CxBR can be found in Gonzalez and Ahlers (1998)[3].

---

3 The interested reader should refer to the ISL website at isl.ucf.edu/CxBR_Appendix_A.doc, where a detailed description of CxBR, with formal definitions, can be found.

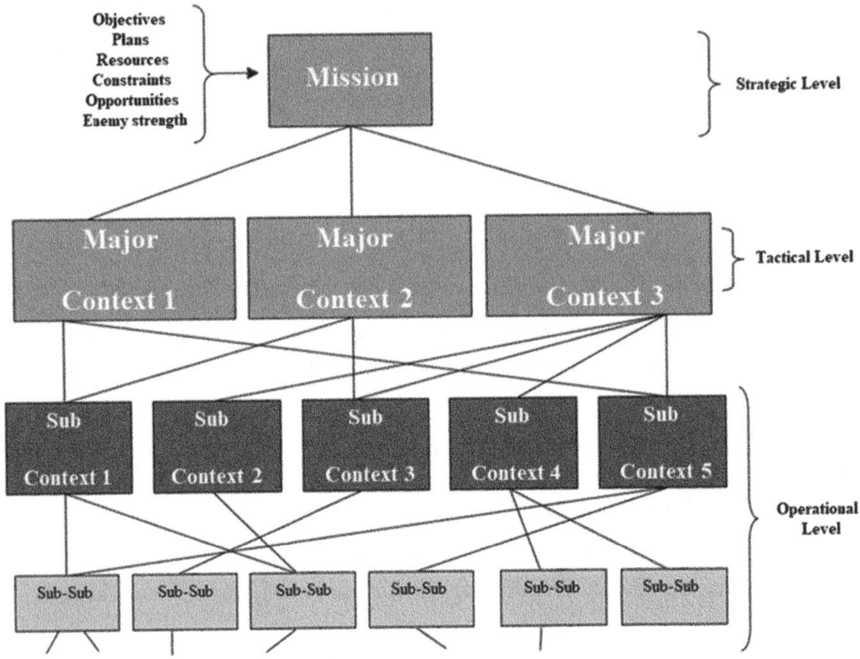

Figure 1.4. *CxBR hierarchical organization for a particular mission. For a color version of this figure, see www.iste.co.uk/brezillon/context.zip*

1.4.4. *Context-mediated behavior (CMB)*

Context-mediated behavior (Turner 1998, 2017; Chapter 5 of this book) is a pragmatic approach to using contextual knowledge that springs from the "scruffy", more cognitively inspired side of AI (see Chapter 3), in particular from case-based reasoning (Kolodner 1993). Context is defined somewhat differently in CMB than in many other approaches. An agent's *situation* is considered to be comprised of all of the features of the world, the agent, the mission, etc., that currently exist. As such, its full extent is in general unknown and unrepresentable in most realistic domains and may, in fact, be infinite (see McCarthy's (1993) view of contexts). In contrast, a *context* in CMB is defined as a class of similar situations that have similar implications for the agent's behavior. Thus, a context represents a generalization of these situations and so refers to the *salient* features of the world, agent, problem, etc., that impact how the agent should behave. In contrast to a situation, a context in CMB is finite and representable.

Contexts are not necessarily monolithic in this view but can often be seen as the merger of other contexts. For example, an AUV may find itself on a search mission in a shallow harbor when there is a storm overhead. It may never have encountered this particular kind of situation before, but it may know about contexts such as "in a storm", "in a harbor", "in shallow water" and "on a search mission". Since the current, novel context can be thought of as a combination of these other known contexts, the agent can combine its knowledge about them to create a representation for the novel context.

CMB is part of a reasoning style called *schema-based reasoning* (Turner 1994), in which most knowledge is stored as packets of related knowledge (*schemas*). *Contextual schemas* (c-schemas) represent contextual knowledge, and *procedural schemas* store knowledge about plans, goal decompositions, scripts, rules, etc. A c-schema contains descriptive and prescriptive knowledge about a context. *Descriptive knowledge* includes facts about the context, both those that must be true for the agent to be considered actually to be in the context as well as predictions about things that may not yet have been encountered by the agent in the current situation. Descriptive knowledge also includes any context-dependent concept meanings. Prescriptive knowledge, on the other hand, tells the agent how it should behave in the context. It includes goal-related knowledge, such as context-dependent goal priorities and which procedural schemas are appropriate to use to achieve goals in the context, and it also includes event-handling knowledge, such as signs that an event may be about to happen, how to diagnose an event given those signs, how important an event is or what will happen if it is ignored, and what to do in response (activate a new goal, perform a particular action, etc.). Prescriptive knowledge also includes "standing orders"[4] for the context, for example, parameter settings, goals to automatically activate, etc.

In CMB, context assessment is modeled from work on abductive medical diagnostic reasoners (Miller et al. 1982). A dynamic memory (Kolodner, 1984; Lawton et al. 1999) watches the evolving situation and provides c-schemas that are "evoked" by it to the context manager, which then does differential diagnosis to determine which one(s) best match the situation. These are merged to represent the context. The process repeats as the situation changes.

CMB is meant to incorporate learning. When used with a dynamic memory, simple inductive learning is automatically supported, as c-schemas are updated as new cases of problem-solving are organized relative to them. Other types of learning can be added to more intelligently modify and create c-schemas, although to date they have not been implemented.

4 Thanks to the late D. Richard Blidberg for this term.

CMB does not make a clear distinction, as does CxG, between contextual knowledge and proceduralized contextual knowledge (PC). All knowledge in the c-schema(s) representing the current context is considered for use by the agent in the context. However, the actual *procedural schemas* (plan-like structures arising from problem-solving in context) used in a situation can be seen as similar to PC. P-schemas, though linked to c-schemas in which they arose or are useful, are stored in their own portion of dynamic memory. When a c-schema suggests one as the way to achieve a goal, the memory is traversed from it as a starting point to find one that fits the context even better. Thus, both the initial p-schema suggested for the context and especially any specialization found can be seen as very similar in spirit to *practices* in CxG.

1.4.5. *Conclusions and lessons learned*

CxBR, CxGs and CMB are very similar in some ways yet still have significant differences. In this section, we look at the similarities and differences, both to explore the properties of pragmatic approaches to contextual reasoning as developed in the approaches, as well as a step toward our ultimate goal of developing a unified approach that incorporates their strengths and avoids their weaknesses.

We base our analysis on the work of Gonzalez and Brézillon (2008) comparing CxG and CxBr, which included identifying dimensions important for tactical behavior, in particular, human tactical behavior. Although these were motivated by the needs of context-based intelligent assistant systems (CIASs), they are general enough to also be important for autonomous real-world AI agents and form the basis of our dimensions, which are:

- basis in human reasoning;
- task-based or goal-based;
- type of formalism;
- scope of application;
- context granularity;
- representation of context dynamics;
- representation of activity;
- environment interface;
- representation of time;

– knowledge acquisition and learning;

– representation of uncertainty and unpredictability;

– provision of explanations.

A summary of the comparison is shown in Table 1.1. In the rest of this section, we discuss the major similarities and differences in more detail.

Criterion	CxBR	CxG	CMB
Basis in human reasoning	Yes	Yes	Cognitively motivated
Task vs. goal based	Task-based	Task-based	Either
Type of formalism	Graphs with multiple parents and shared children	Series–parallel and acyclic directed graphs	Schemas organized along similarities and differences
Scope of application	Hierarchical organization of missions	Real-time modeling of practice development	Hierarchical mission organization
Granularity	Strategic → operational, but esp. tactical level	Tactical → operational	Strategic → operational
Representation of context dynamics	Discrete	Continuous	Discrete
Representation of activity	N/A	At the core of the formalism	Included
Environment interface	Situation-driven	None	Present or absent
Representation of time	Explicit	Implicit	Mostly implicit
Knowledge acquisition and learning	None	Incremental acquisition and learning when system fails	From humans and/or via induction as cases are stored
Uncertainty/unpredictability management	Yes	N/A	Context-specific unanticipated event handling
Explanation provision	None	Practices come from human experts	None

Table 1.1. *Comparison of three pragmatic context-based approaches*

All three approaches generally attempt to emulate human-like intelligence by using context as the basic modeling element. CxG is designed to work with context-based intelligent assistant systems (CAISs), where the agent needs to work like a human and be understandable to a human. CxMB was developed in part to control simulated agents in training simulations, and so it is important that it should also have behavior that is understandable by and similar to humans. CMB has been mostly used for autonomous systems where human-like behavior is not as important as effectively accomplishing the mission, but it is also based on cognitively plausible principles, and in at least two implementations (in a medical diagnostic assistant (Turner 1994) and in a virtual human (Chapter 7)), it needs human-like behavior.

Both CxBR and CxGs are task-based, in sharp contrast to several popular cognitive architectures such as SOAR (Laird et al. 1987), ACT-R (Anderson et al. 1997) and other goal-based approaches that rely on the agent setting a goal, with the system trying to find a production or some other knowledge element that allows the agent to achieve it. In CxG, goals are implicit in the context being experienced. CxBR introduces its mission goals in the separate, non-controlling mission context, of which there is only one. The intermediate goals of the agent in the mission are defined by a plan consisting of a sequence of major contexts, with the transition criteria defined. However, the control contexts themselves contain the functionality to perform the tasks appropriate to that context. In contrast, CMB is more agnostic on the distinction between being goal- or plan-based; it can be either depending on the agent in which it is embedded. If used with a planner, then it will provide context-appropriate goal information, including how to achieve goals in the context; when used with other reasoners (e.g. a deep-learning system), then will provide non-goal-based guidance (e.g. appropriate net topology, weights and hyperparameters).

With respect to formalism, none of CxBR, CxG or CMB are languages, as are, for example, SOAR and ACT-R, nor true representational paradigms, but rather conceptual modeling approaches where knowledge is organized by context. How a model is defined or is actually implemented is unspecified. This gives the developer great flexibility in adapting to the specific needs of the computing environment available. For example, CxBR could easily be modeled through finite state machines, Petri nets or Markov chains. The action knowledge within a major or minor context (sub-context, sub-sub-context, etc.) can likewise be represented through functional programming, rule-based systems, constraint networks or neural networks, among others. Transition knowledge, while traditionally represented by production rules (i.e. *transition rules*), have also been incorporated through neural networks (Stensrud 2005). Problems represented by contextual graphs in CxGs have likewise been implemented in various different ways, such as Petri nets, influence graphs, production rules and others. It is interesting to note that while a contextual graph can be expressed fully in terms of productions, a set of productions cannot be expressed in a unique way as a contextual graph. CMB's contextual schemas can be implemented as frames, semantic networks, description logic, etc., and its schema memory, although intended to be implemented as a dynamic memory, can be anything reasonable. Transitions between contexts have been implemented by an active dynamic memory noticing changes or by predictions directly contained in c-schemas (Chapter 7).

The three approaches differ with respect to the kinds of applications they have been used for. CxBR has been used for the hierarchical (strategic, tactical and operational) organization of missions for real and simulated agents. CxG has been extensively used in modeling practice development for real-world CIASs in several disciplines. CMB has been used in the domains of a medical diagnosis assistant, hierarchical organization of missions for autonomous agents, for organization of multiagent systems and most recently in the domain of context-dependent deep learning.

The approaches can be compared by the level or granularity of reasoning they control: strategic, tactical or operational. CxG exists primarily at the operational level, concerned as it is with practices in particular contexts, and can be at the tactical level. CxBR is primarily at the tactical level, and extends to the strategic and operational levels. CMB similarly extends across all three levels, but makes no hard distinction between contexts existing at different levels; use of a particular contextual schema at one or another of the levels depends on the needs of the reasoner and on its situation. The focus by the three approaches on different levels of granularity promises to allow work that has been done on the strategic and tactical levels (CxBR, CMB) to inform future work at those levels and work focusing on the operational level (CxG) to inform future work at the operational, practice-based level, which was not a focus of the other two.

The approaches can also be compared based on their treatment of the dynamic nature of contexts. All three take context dynamics seriously, with provision for representing the changing context as the situation or the context evolves. CxBR and CMB treat contexts as discrete and relatively large scale. The context is tracked and represented as a sequence of representations of entire contexts, even though any particular context may be represented by several knowledge structures. In contrast, CxG takes a more continuous approach, with the context changing smoothly as the agent progresses through the graph when a CE is instantiated differently.

In summary, the three approaches, developed contemporaneously for the particular needs of their applications, resulted in some important differences but remarkably also many commonalities. This speaks, we feel, to the common needs of pragmatic, context-based reasoning systems.

In particular, we see that all three approaches developed explicit, rich representations of contexts centered around the salient features of similar problem-solving situations and how to behave appropriately in them. Contexts in each are represented as first-class entities, about which the system can reason, not just symbols or a collection of axioms. All approaches also explicitly recognize the changing nature of context and have developed different, though potentially compatible or even synergistic, ways to transition from context to context. All three highlight the benefit of organizing knowledge by context for acquiring knowledge from humans.

In addition, in their differences, the three approaches still point out potentially important characteristics of pragmatic context-based systems. For example, two of the approaches highlight the importance of the tactical and strategic levels of reasoning, while another shows the importance of context in developing and using operational knowledge. Thus, we can see that context plays an important role at all three levels. Two of the approaches highlight the need to acquire context-related knowledge from humans (experts and especially users), while the third shows a route to automatic contextual knowledge acquisition and updates via a process borrowed from memory updates in case-based reasoning.

1.5. Conclusion

Over the last 25 years, technological evolution has relied on a very dynamic development of bottom-up approaches in AI in general, and in context research in particular. This took a more pragmatic, operational form than a formal, top-down one. This is important when trying to solve real-world problems. Any parent who has tried to teach their child how to ride a bike knows the virtue of a pragmatic method (e.g. pushing the child on the bike) rather than explaining to the 5-year-old child the mathematical model of riding a bike (as in the 10-page paper of Fitton and Symons (2018)).

The need to tackle operational knowledge is now acknowledged as requiring contextualized knowledge, by opposition to decontextualized knowledge (e.g. procedures or some formal approaches) that looks to represent general behavior that is robust for any situation. Pragmatic research mainly concerns context-based systems, and it looks to represent how to behave in specific contexts. This is not to say that formal systems are not important. After all, although there were undoubtedly many pragmatic engineering practices used, designing and getting a spacecraft to Mars required a hefty dose of formal methods.

This chapter is situated against the backdrop of the CONTEXT community. This community is international and interdisciplinary precisely because context is a concern found in a number of domains in which humans solve problems. A major objective of the community is sharing results across the domains. The main problem is lack of a unifying model of context, as shown by the large number of definitions found in the literature. This chapter presents successful context-based approaches in different technological domains, and proposes an operational definition of context for at least one of them.

We have found it useful to consider an agent-based approach and to distinguish between data (low-level, objective, agent- and context-independent), information (higher-level, context-dependent, somewhat agent-dependent) and knowledge (high-level, integrated, highly context- and agent-dependent). Based on its contextual knowledge (the part of its knowledge concerning the current focus), a reasoner interprets and uses information, and produces new information. Interpretation, reasoning, decision-making and acting are strongly context-based.

The final contribution of this chapter is an illustration of what a pragmatic, context-based intelligent system looks like by presenting three examples addressing applications in the real world: context-based reasoning (CxBR), contextual graphs (CxG) and context-mediated behavior (CMB). These have been used in different domains (mission management for the army for CxBR; medical diagnosis and autonomous agents for CMB; and several different domains for CxG). We show that they share many commonalities and, though there are differences, there are no incompatibilities. They all put context front stage for their use of data, information, knowledge and reasoning in a uniform representation. They use operational knowledge as previous approaches used formal knowledge. The main interest of operational knowledge is to adapt formal knowledge to each situation in which it is used and for each specific problem to which it is applied. This thus leads to a contextualization of the reasoning and to pragmatic, context-based intelligent (assistant or autonomous) systems, which is the subject of Chapter 3 of this book. The comparison leaves us hopeful that an integrated approach can be devised that combines the strengths of the approaches while avoiding their weaknesses.

1.6. References

Anderson, J.R., Matessa, M., Lebiere, C. (1997). ACT-R: A theory of higher level cognition and its relation to visual attention. *Human–Computer Interaction*, 12(4), 439–462.

Bazire, M. and Brézillon, P. (2005). Understanding context before using it. In *Modeling and Using Context: Proceedings of the International and Interdisciplinary Conference*, Dey, A., Kokinov, B., Leake, D., Turner, R. (eds). Springer Verlag, Berlin.

Brézillon, P. (1989). Interpretation and rule packet in expert systems. Application to the SEPT expert system. *Proceedings of the International Conference on Knowledge Based Computer Systems*, 78–87, Bombay, India.

Brézillon, P. and Abu-Hakima, S. (1995). Using knowledge in its context: Report on the IJCAI-93 Workshop. *AI Magazine*, 16(1), 87–91.

Brézillon, P., Pasquier, L., Pomerol, J.-C. (2000). Representing operational knowledge by contextual graphs. *Advances in Artificial Intelligence*, 245–258.

Chandrasekaran, B., Gomez, F., Mittal, S., Smith, J. (1979). An approach to medical diagnosis based on conceptual structures. *Proceedings of the Sixth International Joint Conference on Artificial Intelligence*. Tokyo, Japan.

Fitton, B. and Symons, D. (2018). A mathematical model for simulating cycling: Applied to track cycling. *Sport Engineering*, 21, 409–418.

Gonzalez, A.J. and Ahlers, R. (1998). Context-based representation of intelligent behavior in training simulations. *Transactions of the Society of Computer Simulation*, 15(4), 153–166.

Gonzalez, A.J. and Brézillon, P. (2008). Comparing two context-driven approaches for representation of human tactical behavior. *Knowledge Engineering Review*, 23(2), 1–21.

Gonzalez, A.J., Stensrud, B.S., Barrett, G. (2008). Formalizing context-based reasoning: A modeling paradigm for representing tactical human behavior. *International Journal of Intelligent Systems*, 23(7), 822–847.

Kolodner, J.L. (1984). *Retrieval and Organizational Strategies in Conceptual Memory*. Lawrence Erlbaum Associates, Mahwah.

Kolodner, J.L. (1993). *Case-Based Reasoning*. Morgan Kaufman, Burlington.

Kuhn, T.S. (1970). *The Structure of Scientific Revolutions*, 2nd edition. University of Chicago Press, Chicago.

Laird, J.E., Newell, A., Rosenbloom, P.S. (1987). SOAR: An architecture for general intelligence. *Artificial Intelligence*, 33(1), 1–64.

Lawton, J.H., Turner, R.M., Turner, E.H. (1999). A unified long-term memory system. *Proceedings of the Third International Conference on Case-Based Reasoning*. Seeon Monastery, Germany.

McCarthy, J. (1993). Notes on formalizing context. *Proceedings of the 13th International Joint Conference on Artificial Intelligence*, 555–560, Chambery, France.

McCarthy, J. and Hayes, P.J. (1969). Some philosophical problems from the standpoint of artificial intelligence. In *Machine Intelligence 4*, Meltzer, B. and Michie, D. (eds). Edinburgh University Press, Edinburgh.

Miller, R.A., Pople, H.E., Myers, J.D. (1982). INTERNIST – 1, an experimental computer-based diagnostic consultant for general internal medicine. *New England Journal of Medicine*, 307, 468–476.

Newell, A. and Simon, H.A. (1972). *Human Problem Solving*. Prentice Hall, Englewood Cliffs.

Pasquier, L., Brézillon, P., Pomerol, J.-C. (2000). From representation of operational knowledge to practical decision making in operations. *Decision Support Through Knowledge Management*, 301–320.

Russell, S. and Norvig, P. (2020). *Artificial Intelligence: A Modern Approach*, 4th edition. Pearson, London.

Simon, H.A. (1960). *The New Science of Management Decision*. Harper & Brothers, New York.

Stensrud, B.S. (2005). FAMTILE: An algorithm for learning high-level tactical behavior from observation. Doctoral Dissertation, University of Central Florida, Orlando, USA.

Turner, R.M. (1994). *Adaptive Reasoning for Real-World Problems: A Schema-Based Approach*. Lawrence Erlbaum Associates, Mahwah.

Turner, R.M. (1998). Context-mediated behavior for intelligent agents. *International Journal of Human–Computer Studies*, 48(3), 307–330.

Turner, R.M. (2017). Context-mediated behavior. *Modeling and Using Context*, 17(1).

2

Modeling and Using Context: 25 Years of Lessons Learned

This chapter presents the results of 25 years of research on modeling and using context in real-world applications. The first milestone was an operational definition of context relating it to a focus (problem-solving, decision-making, task realization). The focus provides context with links between context and knowledge, a division between contextual knowledge and external knowledge with respect to a focus, and a dynamic linked to reasoning evolution. Conversely, context makes focus more explicit. The second milestone was a context-based representation of decision trees as contextual graphs with contextual elements (expressing the focus) instead of chance or event nodes. The operational definition of context opens up a conceptual framework that can be associated with an implementation framework. The third milestone was the development of a context-based representation formalism that led to the contextual graphs (CxG) software that has been used in about 20 real-world applications. The fourth milestone concerns an extension of the CxG representation formalism for representing group activity with cyclic use of a CxG for simulation purposes. The formalism now allows modeling of "nonlinear" behaviors of a group (e.g. interaction, negotiation, etc.) as well as of a user (e.g. checking alternatives). The fifth milestone opens the way to context-based intelligent assistant systems.

2.1. Introduction

We started research on context through two projects (SEPT and SART) within the technological domain in which exist formal models of equipment to monitor. In the first project, SEPT with *Electricity de France* (Bau and Brézillon 1992), we rapidly discovered that having a formal model was not sufficient because a number

---

Chapter written by Patrick BRÉZILLON.

of parameters were not controllable or not considered in the formal model, although they intervene indirectly in it. For example, equipment in an extra high voltage (EHV) substation in a narrow valley between mountains cannot function as if it were on a plain. More globally (the observation was reinforced in the SART project (Brézillon et al. 2000) for the subway in Paris and Rio de Janeiro), if it is possible to formally model the functioning of equipment in an EHV substation, some external elements have to be considered: the actor, the task the actor must realize (e.g. monitoring of equipment functioning in the substation), the situation in which the task is realized, and the local environment where there are resources available for the task. All of these external elements may constrain the normal functioning of what is in the focus. Actor, task, situation and environment are sources of contextual knowledge that matters for a focus.

As a result, context is strongly related to knowledge and reasoning, both in models and in the heads of users. Working on real-world applications obliged us to avoid arbitrary simplification in modeling context, and our interest in decision-making led us to establish an operational definition of context. It was then possible to really model context in terms of contextual elements that had a concrete meaning for actors with whom we worked across a very large spectrum of domains[1]: power systems, subway exploitation, enology with winemaking, computer security, collaborative understanding, car driver support, website analysis, self-training of car drivers, medical image access, contextualization of platforms of interoperable open source tools for enterprises, database administration, a strategic infrastructure project in transportation, a cognition-driven explorer for histopathology for breast cancer grading, processing and sharing of medical images, and contextualized support on a tablet for the transmission of command information.

The operational definition of context and the in-depth study of different aspects of context have allowed the design and implementation of a context-based representation formalism called contextual graphs (CxG). The CxG formalism has been used with success in all the applications cited in the previous paragraph. In this chapter, we only provide the main features of applications needed to understand some lessons learned about context and its role in applications; most of the key references on the CxG formalism are given in Brézillon (2018).

In the first two applications, the unique consideration of a task model (e.g. technical manuals in the SEPT project and incident reports in the SART project) was insufficient for developing a competent computer software: it is important also to

1 The list of the publications (and PDF for some of them) on most of all the projects and applications is given at http://www-poleia.lip6.fr/~brezil/Pages2/Publications/index.html.

consider the way in which the task is realized by an actor. In the SART project, operators give great importance to contextual knowledge for monitoring subway lines during an incident (e.g. at rush hour). An operator said, "when I am informed of an incident on my line, I first look to the current state of the whole line to first gather contextual information", and then make decision accordingly. Landauer and Bellman (1993) similarly conclude that the process of problem-solving involves three major steps: find_context, pose_problem and study_problem. The context includes conditions on the existence of certain external information and partial mappings between elements.

Operators interpret the task model (company procedures that are supposed to be mandatory) according to the contextual information collected to determine how to solve the incident with minimal cost, at policy and strategic levels (e.g. prioritizing a quick solution to avoid a degraded situation in the SEPT project, and traveler security and traffic regularity in the SART project). Again, operators use contextual knowledge from sources like the task, the situation and the local environment.

The remainder of this chapter is organized into three parts. Section 2.2 presents knowledge in action to point out the relationships between operational knowledge and (1) contextual knowledge and (2) mental models of actors. This leads to a discussion of the modeling of operational knowledge and, if it is not possible, the contextual alternative. Section 2.3 presents context in action through its conceptual modeling under different aspects and the implementation in the contextual graphs (CxG) representation formalism. Section 2.4 develops the action of context in real-world applications, first with respect to a precise focus (task realization, decision-making, problem-solving), and second with respect to actors responsible for a focus alone or in a group. The section ends with a discussion of the extension to the CxG formalism for modeling group activity. Section 2.5 sums up the lessons learned and the results obtained over these 25 years.

2.2. Knowledge in action

2.2.1. *Operational knowledge and contextual knowledge*

Brézillon and Pomerol (2001) said that context is knowledge about the instantiation of *know how*. It is the framework that reveals *know how*. In other words, knowledge is the information that is integrated and understood in the mind of an actor, and information results from data interpretation. The focus allows two parts of context to be differentiated, namely, contextual knowledge and external knowledge.

Contextual knowledge constitutes a set of elements more or less related to a focus in a flat way. *External knowledge* concerns elements that are not important for the focus at the moment we consider it. The frontier between contextual and external knowledge is porous, and elements move between contextual knowledge and external knowledge as the focus progresses. Selection of the elements of the contextual knowledge concerning the focus is a kind of first instantiation of the contextual knowledge or background context that needs some further specification to perfectly fit the task at hand. The precision and differentiation brought to the contextual knowledge is also part of the proceduralization process. “Knowing how” is “instantiating in doing”, and contextual knowledge obeys a dynamic of instantiation and proceduralization during action. The important issue is the passage from contextual knowledge to *proceduralized context*, the part of context used in the focus. The proceduralization aspect corresponds to the instantiation of the contextual elements entering in the focus (see also Grimshaw et al. 2000).

Contextual knowledge is personal to an actor and is evoked by situations and events. It is something that is stored in long-term memory, and recalled as a whole at a certain step of task realization. Such a conceptual framework supposes a uniform representation of the elements of knowledge, reasoning and context. The actor’s focus is guided by a mental representation that the actor has of the task realization in its current step, the situation in which the task is being realized and the local environment in which resources are available. The mental model of the actor represents the personal context in which the actor interprets the focus. It describes something important about knowledge that is not captured in traditional knowledge measures, such as its operational and dynamic organization. Knowledge organization is different from one actor to another.

2.2.2. *Operational knowledge and mental models*

In a real-world application, an expert develops a mental model based on their experience accumulated from task realizations (use of domain knowledge in different contexts). Experience relies on operational knowledge as well as domain knowledge.

In the TACTIC project (Kabil et al. 2015), experts were asked to develop mental maps as a semi-structured expression of their mental models (this was also done in another project in medicine, Brézillon et al. (2014)). Their mental maps show depth- and breadth-first strategies for modeling how they realized their task. Usually, reasoning is between a “depth-first” strategy and a “breadth-first” strategy. The "depth-first" strategy goes to the finest possible granularity on a line of reasoning in order to anticipate as much as possible the course of events from a maximum of contextual elements and their instantiations. It assumes that we know what to do and how to get there quickly. This strategy makes it possible to study the technical feasibility of an approach as well as the needs in terms of resources, and, in a second step, gradually expands this approach. Conversely, the "breadth-first" strategy is applied when it is necessary to first consider all possible options before making decisions. Mental models in a breadth-first strategy are interesting for also detecting, for example, weak signals because experts keep an "open mind". Their focus is not on task realization in an isolated way but rather on how the realization is placed in its context. Such experts reason at a strategic level and do not consider low-level details of task realization. Conversely, maps made by experts in a depth-first strategy can be used to decide quickly at the tactical level.

The breadth-first strategy is observed in the maps of experts that maintain the more relevant contextual elements, even if not directly necessary in the task realization. For example, "environment" was considered by an expert to be only a part of "situation", even if environment is the main source of contextual elements on the battlefield map. The reason is that experts consider environment through what they really need to know for realizing their task. For example, it is only when there is contact with the enemy that participants zoom in on the location of the enemy (e.g. in a city or landscape). The participants are thus more interested in an external event due to its effect on task realization rather than because of its origin.

The interplay between depth- and breadth-first strategies is similar to global search (FlexMIm project; Brézillon et al. (2014)). Both are based on a cycle beginning with a phase of rapid exploration for finding zones of interest (global search) followed by a phase of precise analysis of the zones ultimately judged as being of interest (local search). Context-aware systems also distinguish two types of context: the "local" context that is close to the focus and highly detailed, and the "distant" context that is general (with fewer details). In the spirit of Sowa's (2000) conceptual graphs, all of the concepts can be expanded in the local context and aggregated in the distant context. Local details are needed for the local interactions with a structure. The global context is needed to tell the actor what other parts of the structure exist and where they are. Global information is important even in the simple interpretation of local details.

2.2.3. *Modeling operational knowledge*

The two types of reasoning (breadth-first strategy and depth-first strategy) are shown in Figures 2.1 and 2.2, respectively (only the general form of the two maps is important here). The mental model of the task "give a reconnoitering order" (find out what the enemy is doing) was developed by knowledge acquisition techniques from a panel of 10 experts that routinely used a battlefield simulator for different purposes. Each expert had expressed in a mind map the knowledge they generally used for realizing the task and simplified the map for the specific simulation (i.e. in a specific context). Figure 2.1 shows the map made by a tactical expert (a simulator trainer) and Figure 2.2 shows the map of an operational expert (a developer). Gray items correspond to the contextual elements not used in the specific scenario on which they had to work.

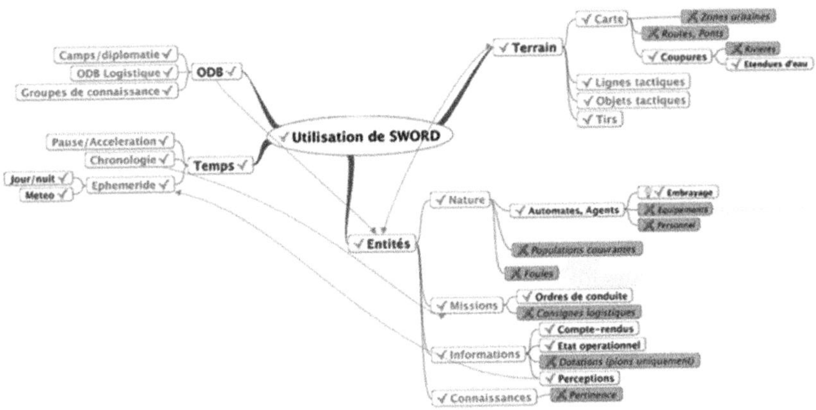

Figure 2.1. *Knowledge organization from a tactical view (Brézillon 2018). For a color version of this figure, see www.iste.co.uk/brezillon/context.zip*

Figure 2.1 shows the knowledge organization of the tactical expert (the general tree), which is very close to that of the operator interacting with the simulator (the end-user). The focus is limited to the position of the mission in the context of the battlefield and lets all other problems remain at the periphery of the mental model. The expert also introduces links between leaves in the mind map to express their strong compilation of the contextual elements. For example, the item "perceptions" is linked to the item "ephemeris" because the useful information on weather and day/night only addresses the question of the soldier's vision around them.

Figure 2.2 shows how the operational expert organizes (the general tree) their knowledge in depth for analyzing more what the execution of an action on the simulator is than for analyzing what the action is on the battlefield (i.e. in the simulation).

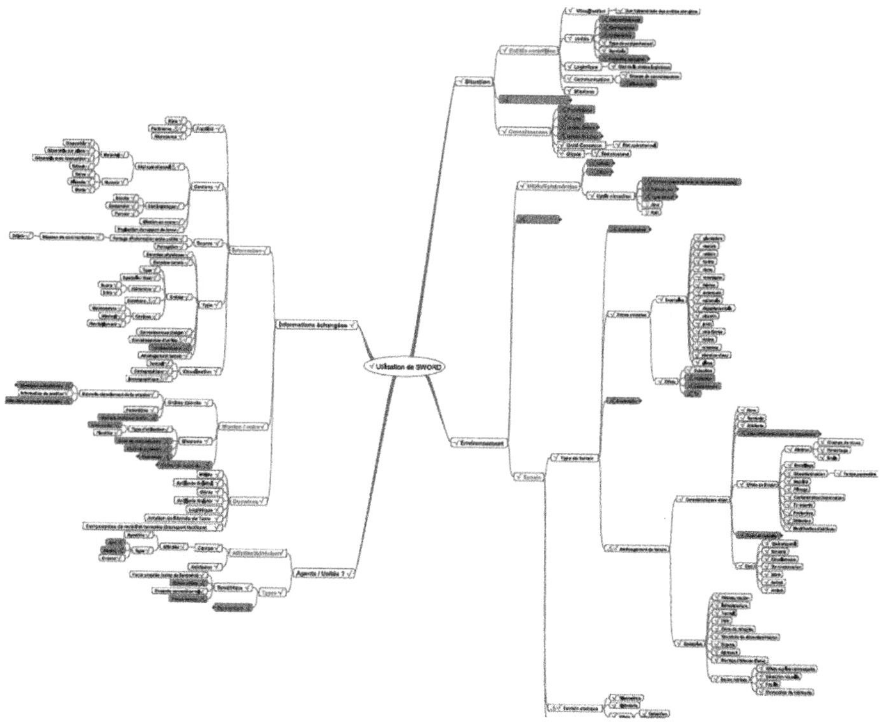

Figure 2.2. *Knowledge organization from an operational view (Brézillon 2018). For a color version of this figure, see www.iste.co.uk/brezillon/context.zip*

2.2.4. *Indirect modeling from experience reuse*

The direct reuse of known practices instead of starting from scratch with a procedure is observed in domains where experts' experience is preferred to domain knowledge that is weakly formalized, if at all. Experts rely on highly compiled experience, and the reuse of their experience is never direct because it was obtained in numerous specific contexts. Thus, reusing a past experience implies that the actor follows these three steps:

1) "contextualize" (i.e. find the contextualization of) a known past experience in relation to the experience at hand;

2) "decontextualize" the past experience for finding the model (i.e. retrieving the contextual elements from their instantiations);

3) "recontextualize" the model of the past experience by re-instantiating its contextual elements in the context at hand.

This process of contextualization–decontextualization–recontextualization (CDR) is used in different works.

Brannen (2004) analyzes the first two attempts the Walt Disney Company made at internationalization: Tokyo Disneyland (1983) and Disneyland Paris (1992). The goal was to implant a strict copy of the initial park at Anaheim (USA) in Japan and in France, and the internationalization challenge was assessing the fit of what they wished to transfer abroad with the new host environment. Generally, firms transfer their whole organizations, including business models, steeped in image-based, ideological, people-dependent management that are even more closely related to context and sociocultural environment in which they are enacted. However, if the source is significantly foreign from the target, the transferred assets may not fit the target context in the host country. This was the case of Disneyland Paris, but after the positive experience with Tokyo Disneyland (no CDR needed), the context of the French target was too different (recontextualization was mandatory, as discovered several years later).

Fan et al. (2011) applied the CDR process in the search for a scientific workflow (SWF) in virtual screening of the H5N1 virus (see Figure 2.3): a researcher extracts a successful SWF from a repository that they regard as close to their focus (phase of contextualization), extracts the corresponding SWF model (phase of decontextualization) and finally instantiates the SWF model in the working context at hand (phase of recontextualization). Then, a workflow engine ensures the execution phase in which input data is consumed according to the instantiated SWF, and output data are produced as results. If it is a success, the new scientific workflow is put in the repository for other scientists. Otherwise, a new iteration is needed by reconsidering either the phase of decontextualization for finding a new model, or by trying a different recontextualization of a new scientific workflow.

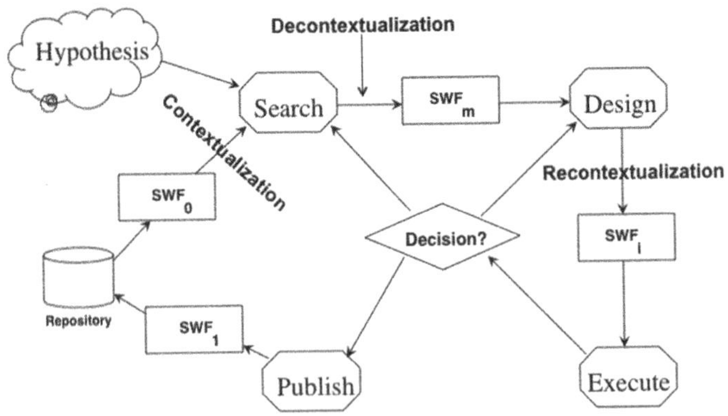

Figure 2.3. *Flexibility inside a workflow lifecycle (from Fan et al. 2011)*

The experience acquired by the researcher during this process relies on context management. This modeling allows context, and consequently operators' behavior, to be made explicit.

2.2.5. *Lessons learned*

The main lessons learned were the following.

– Tactical experts tend to create contextual graphs with a predominant parallel structure (breadth-first strategy), while operational experts are more prone to model their problems by developing contextual graphs with a predominant series structure (depth-first strategy). The main point is that the former reasons on the basis of their experience (i.e. structured knowledge), while the latter assembles contextual elements in more of a reactive way rather than in a pro-active one.

– The quality of the task-realization model depends directly on the expert's knowledge about the task realization, the situation and the local environment. The operational expert focuses on the narrow understanding of the problem, while a tactical expert sees the problem at hand among other problems to find a solution.

– The tactical expert thinks in an organized and structured way, avoids assumptions leading to unrealistic scenarios and identifies actions that can be performed in parallel. The operational expert executes actions sequentially without paying attention to the logic of the task. The operational expert may add sub-tasks or change the order of the subtasks, while the tactical expert applies the general scenario of first collecting contextual information and second applying the better strategy for the proceduralized context found.

– The tactical expert has a holistic view of the task realization and discriminates relevant from useless contextual elements in a kind of "cognitive simulation" to look ahead in the task realization. The operational expert has a fragmented view of the task realization and, thus, reasons reactively, discovering the task during its realization, sometimes missing some details, considering irrelevant aspects or establishing no links between the different steps.

– An expert map, as in Figures 2.1 and 2.2, expresses what could be a working context, with its contextual elements, their possible instantiations and their organization. Studying the expert maps shows there is no unique model of the working context (see the difference between the expert maps in Figures 2.1 and 2.2). Clearly, the organization of the contextual elements depends on the experience of the actor that must realize the task. The tactical expert is not interested in the details and retains only the direction to follow. Conversely, the operational expert follows one direction and precisely analyzes what to do.

2.3. Context in action

2.3.1. *Conceptual modeling*

Brézillon and Pomerol (1999) defined context as "what constrains the focus without intervening in it explicitly". The focus could be decision-making, problem-solving, or task realization. This definition implies two consequences: first, we cannot speak of context in an abstract way – it is always relative to a focus; and second, context and focus are interdependent – the context makes the focus explicit and the focus defines the relevant contextual knowledge needed at each step. Making context explicit implies that our modeling of a task concerns its realization as a process and not just a formal description of the task model. From an implementational viewpoint, context is an aggregation of contextual elements coming from four sources, namely the user, the task, the situation and the local environment.

The first key point of context modeling is the association of conceptual and operational views, ensuring direct implementation of the conceptual model of context. Figure 2.4 shows our approach from the conceptual view to the operational view. This double viewpoint (conceptual and operational frameworks) constitutes the basis of all of our (successful) research in real-world applications.

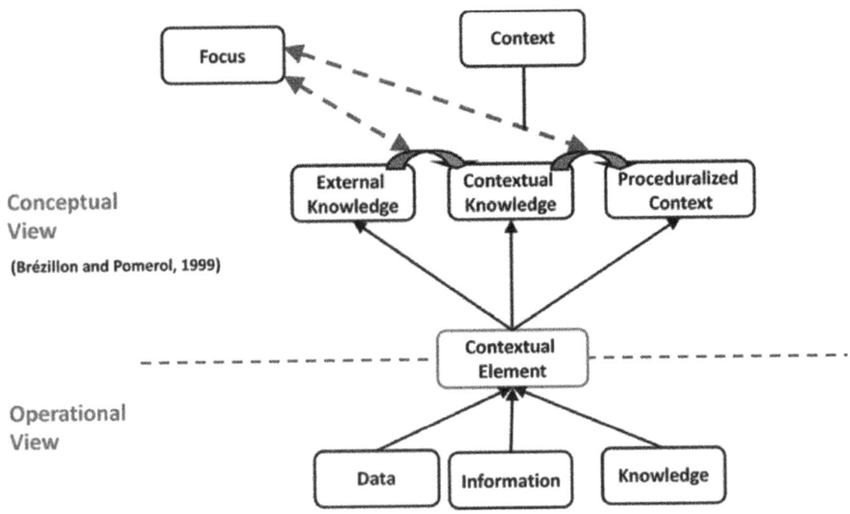

Figure 2.4. *Conceptual and operational views on contextual elements*

2.3.2. *A typology of contexts*

The *working context* contains the contextual elements that an actor associates with their focus (e.g. task realization), all the known values of each contextual element and a chosen instantiation. Arriving at a contextual element, the practice development needs its instantiation to select the right path to follow and the action to execute. The instantiation can be known prior to the task realization, provided by the actor when (and if) needed, or found by a system using the contextual graph in the local environment.

The *shared context* concerns a task realized by a group of actors. The shared context is developed from the working contexts of the group members. It evolves continuously according to the group focus and the addition of contextual elements by group members. A group task realization starts by establishing the shared context. The group members first gather the maximum number of contextual elements without looking for a global picture (in a kind of brainstorming session), a member proposes a contextual element that is supposedly relevant and other actors accept it or ask for an explanation. To accept the new contextual element, all group members must be able to relate it to their individual working contexts. Thus, joint development of the shared context allows group members to keep their mental models compatible (Karsenty and Brézillon 1995). An example of collaborative building of a proceduralized context is shown in an application presented in section 2.3.3.3.

The phase of shared-context building does not only concern contextual knowledge but also external knowledge that the proposal originator has (e.g. "I had the opportunity to visit a place concerned with this topic a couple of years ago"). When group members reach a consensus on their shared context, they enter the second phase of the process by organizing, assembling and structuring most of the contextual elements of their shared context in a proceduralized context, allowing us to reach a group focus. In the SART project, operators work eight hours a day in the control room of a subway line and are replaced by other operators who receive a debriefing on what has happened before their arrival and the difficulties encountered. This sharing of contextual information between actors is essential for assuring continuity of the normal operation of the subway line (and indeed in other places). It is also the case for several projects we worked on after the SART project. Group activity is mandatory when either the task model is too complex for a single actor (e.g. monitoring a nuclear plant) or a real task model does not exist (e.g. in medicine and in domains such as winemaking). In some domains, such as scientific workflows, actors have no task model at all and are obliged to adopt the contextualization–decontextualization–recontextualization approach.

A *proceduralized context* is quite similar, in spirit at least, to a chunk of knowledge (Miller 1956; Schank 1982) and, in its creation, to Clancey's (2002) view of diagnosis as the construction of a situation-specific model. A proceduralized context is built progressively with practice development. This “chunk of contextual knowledge” is activated and structured to facilitate decision-making or task realization (Pomerol and Brézillon 1999). Concretely, the proceduralized context is the ordered series of contextual elements that are instantiated during practice development. If contextual knowledge is background knowledge for the focus, the proceduralized context is foreground, that is, immediately useful at the given step of the focus.

It is possible to retrieve how the proceduralized context was built, the reasons behind choices (contextual elements considered and their instantiations), the alternatives abandoned (actions corresponding to the values of the contextual elements not retained), etc. A proceduralized context is either static, when related to an item, or dynamic, when related to the development of the practice (see section 2.4).

2.3.3. *About contextual elements*

2.3.3.1. *Organization around the focus*

The onion metaphor was proposed in the SART project (Brézillon et al. 1997, 2000) to illustrate context organization. The focus (the successive steps in incident solving) is the onion heart, and we observed that contextual elements were organized in layers around the heart. Figure 2.5 shows the solving of the incident “sick traveler in a train”. The step “stop at the next station” is the onion heart, and dotted lines around it represent the successive layers of contextual elements. For instance, the contextual element “procedures” is in the first layer, “past experience” is in the second, which is an explanation for procedure, etc.

The lessons learned with the onion metaphor reveal several interesting points for our modeling of context for real-world applications.

– Only contextual knowledge is considered, which avoids the problem of the infinite dimension noted by McCarthy (1993), linked to external knowledge.

– A step has a meaning within a context. Contextual knowledge does not intervene explicitly at this step but constrains it.

– Each contextual element itself has meaning within a context. For example, the contextual element “procedures” in Figure 2.5 has its own context with contextual elements like “past experience” and “do not touch an injured traveler”. It is an illustration of McCarthy’s (1993) claim about the definition of a context in an outer context.

– Layers imply the partial ordering of contextual elements with respect to the focus. For example, “procedures” (layer 1) < “don’t touch an injured traveler” (layer 2) < “driver not trained to treat a sick person” (layer 3) <, etc.

– Incidents may be linked to one another through a sequence of ordered contextual elements. For example, “sick traveler in a train” is linked to the incident “traveler on railway” through the contextual elements “procedures”, “ensure traveler’s safety”, “avoid stopping in a tunnel”, “problem with traveler’s behavior”, this latter contextual element being shared with the incident “traveler on railway”.

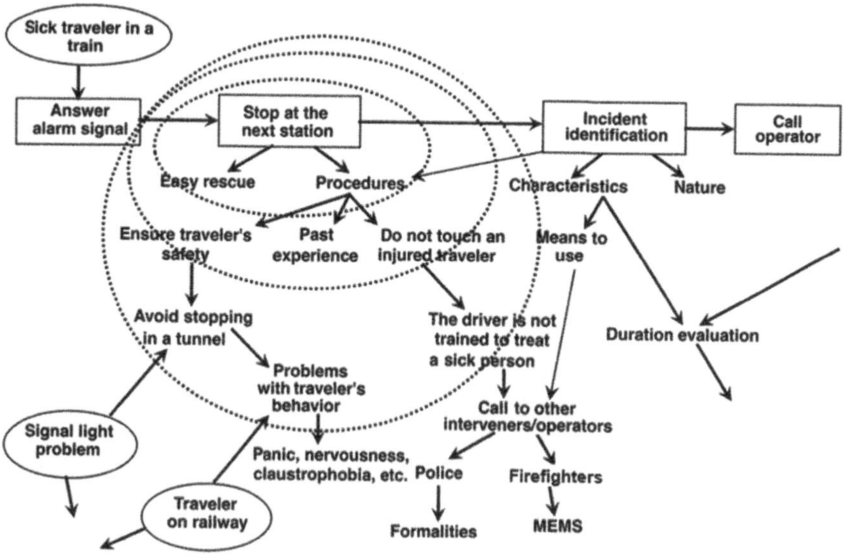

Figure 2.5. *Contextual organization for the incident “sick traveler in a train”*

The onion metaphor allows the development of explanations of different natures and at different levels of detail that correspond to the layers. It is useful in many different applications. In the wine-making application (Agabra et al. 1997), “far” contextual elements may be more crucial for task realization than some close ones. For example, it is well known (in France) that the contextual element “weather” a few days before vintage – which is qualitatively far from “wine fermentation” – plays a central role in winemaking. It is particularly important because, unlike the technical domain, often the task model may be too difficult to model formally (see section 2.2.4). For example, enologists regard the wine fermentation step as a black box due to numerous and complex chemical reactions that occur at this point.

The onion metaphor shows a representation of contextual elements that enables us to remember that even a distant contextual element can be more important than a close one. Conversely, fisheye views (Furnas 1986; Pook et al. 2000) are also one way of integrating context and focus into a single view on local details and global context simultaneously. However, the global context cannot be exploited as local details.

2.3.3.2. *Instantiation of the contextual elements*

The second key point of the context representation is the distinction between a contextual element and its instantiation, a distinction often not considered in other representation formalisms. The explicit introduction of contextual elements for representing contextual knowledge and the distinction with their values offer new perspectives: the acquisition of a new practice corresponds to the addition of either a new contextual element (learning by accommodation) or a new value for an existing contextual element (learning by assimilation), the latter case only leading to refinement of the practice. Several types of values of a contextual element are possible, from qualitative to quantitative. For example, the contextual element "Temperature?" may have values like "warm" or "cold" but, according to the focus, quantitative values like "24.5°C" are possible instead of "warm".

During practice development, the instantiation of contextual elements may be altered by either an external or an internal event. The external event corresponds to an unpredicted event (e.g. the power supply being cut off), and thus is not represented in the task realization. An internal event occurs as the result of an action execution. For example, a fuse becomes too hot and blows. The change of instantiation of a contextual element modifies the working context, and may lead to a change of task to realize if the initial hypothesis is abandoned, and to consider the development of another practice in the task realization, or to modify the conclusion of the task realization.

2.3.3.3. *Collaborative building of the proceduralized context*

Figure 2.6 represents the building of a proceduralized context by two actors for a collective focus. Actor_1 introduces the contextual elements CE_1 from their working context into the shared context. Because CE_1 belongs to Actor_2's working context too, the contextual element is accepted. When Actor_1 proposes the introduction of CE_2, Actor_2, which has no reference to it in their working context, needs explanations to accept CE_2 and both agree on CE_2' (compatible interpretation of CE_2 in the shared context). To accept CE_2', the receiver builds an interpretation CE_2" in their working context. Once Actor_2 agrees to the contextual element, the contextual elements of the shared context are combined into a proceduralized context for addressing the focus.

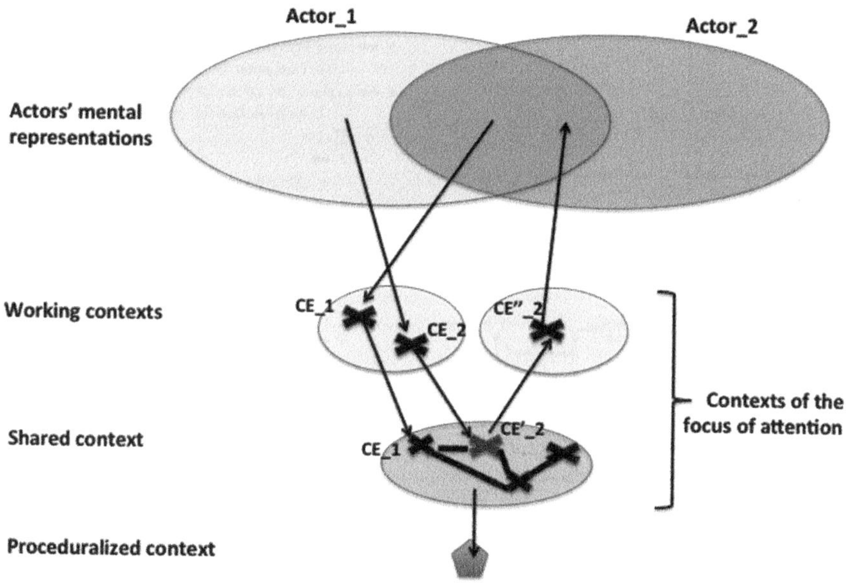

Figure 2.6. *Collaborative building of a proceduralized context. For a color version of this figure, see www.iste.co.uk/brezillon/context.zip*

Brézillon et al. (2006) describe an experiment for modeling verbal exchanges between two actors for collaboratively generating answers to questions like "How does the oyster make pearls?" and "How does water become mineral?". The collaborative process of answer building includes a first phase of building the shared context of the collaboration. Each actor proposes elements of their working context. The other actor may agree or ask for an explanation and eventually negotiates with the first actor. Finally, actors agree on a shared context that satisfies both of them. In a second phase, the actors organize, assemble and structure the contextual elements of the shared context to generate their proceduralized context that will be integrated into the answer building.

Some results of the project are as follows. Eleven pairs of participants had to address 16 questions (176 MP3 files of 1 minute 30 seconds correspond to an answer). The experiment setup was in two phases: a collaboration phase (1m 30s) and then a reading/comprehension phase, including analysis of eye movements and answers to questions. Here, we only discuss the first phase. Participants naturally followed a process in four steps, namely, reformulate the question for understanding, find an example to confirm their understanding, gather domain knowledge and elements of their working context, and build the answer either by looking for characteristics or by assembling explanation elements (for an integration).

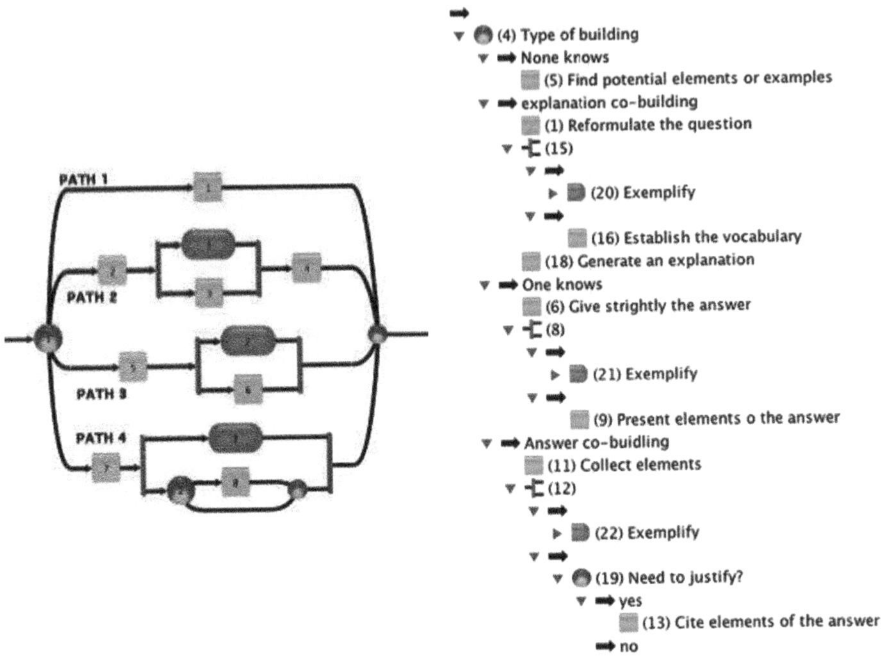

Figure 2.7. *Contextual graph of the identified collaborative answer building processes. For a color version of this figure, see www.iste.co.uk/brezillon/context.zip*

Figure 2.7 represents in the CxG formalism (for an explanation of the components, see section 2.3.4) the four types of the collaboration model found.

Path 1: no knowledge about the answer;

Path 2: no answer but elements of explanation;

Path 3: two-way knowledge (co-building of the answer);

Path 4: one-way knowledge (answer provided by only one participant).

Several observations could be made from the elicitation phase.

– Subjects had a problem finding the right level of granularity for their answer. The answer can be given at different levels of granularity, and can be at the right level as well as at too fine a granularity (too many details) or too coarse a granularity (rough description of the answer), such as “gas” instead of “CO_2”.

– Subjects often immediately repeat the question to be sure they correctly understand it, that is, to be able to find some relationships between elements of the questions and elements of answers from their personal working contexts.

– Subjects first gather contextual information, not to immediately build the answer but to determine the right granularity of their answer, and second to assemble the instantiated contextual elements to use in the proceduralized context. When subjects cannot identify the right level of granularity, they enter into a process of explanation to justify their proceduralized context.

– Subjects can know the answer without knowing elements of the answer. As a result, subjects express an external and superficial viewpoint.

– Collaboration is the minimal expression of cooperation: one subject leads the interaction and the other subject only feeds in information (or only agrees), reinforcing the statements of the other.

– Subjects provide explanations spontaneously to (1) strengthen a known answer, (2) make progress in the co-construction of the answer and (3) be sure of the granularity of the answer. The explanation is generally less precise than the answer (i.e. generally at too general a level), but useful between them to reinforce their ties.

– Several groups of subjects were confused and explained instead of giving the answer (thus with additional details not necessary).

The shared context contains all the contextual knowledge discussed and accepted by all the actors during this process of shared context building. Then, actors build the proceduralized context. Once the focus is addressed, the proceduralized context enters actors' mental models. Later, this proceduralized context may be recalled as a whole for another task realization. Thus, the more an actor experiments, the more the actor acquires structured knowledge, and the more actors work together, the more they reinforce their ties too.

2.3.4. *Implementation of the contextual graphs formalism*

Based on the notion of the contextual element, we have developed a formalism of representation called contextual graphs (CxG) (Brézillon et al. 2000). This formalism relies on four components, namely actions, contextual elements, activities and ESIA (executive structure of independent activities), and has been implemented as a CxG software. Figure 2.7 shows an example of use of the CxG software. Actions are green square boxes, contextual elements are represented by a contextual node (blue circle with a number) and a recombination node (blue circle without number) where end the paths coming from the contextual nodes, activities (contextual subgraph identified by actors as independent of the main graph) are represented by pink ovals. An ESIA is represented by two vertical red bars with independent activities between them. The independent activities can be treated in any order, but all will be executed before to resume the traversal of the main graph. This allows us to reduce the size of contextual graph; otherwise, it would be necessary to have twice the same subtree after a branch activity1–activity2 and another branch activity2–activity1.

A contextual element in the CxG formalism allows management of alternatives on specific parts of practices according to their specific instantiations (i.e. the values of the focus at hand). A contextual element is represented by a pair {contextual node, recombination node}, and the legend is a question about its instantiation (e.g. "Temperature?"). Between the contextual and recombination nodes, there are as many exclusive branches as known practices (and instantiations, for example, "warm" and "cold"). A value (quantitative or qualitative) of the contextual element corresponds to a branch between the two nodes with an action or an activity or an ESIA specific to the decision-making on this instantiated contextual element. When the inference engine arrives at a contextual node, it needs to instantiate the contextual element at one of its values (known initially or asked of the user) to select the corresponding branch and progresses in the development of the practice. At the recombination node, the instantiation does not matter anymore, since the corresponding action has been completed. The sequence of instantiations during practice development corresponds to a context-specific modeling of the practice (the proceduralized context), while contextual elements belong to the model of the task realization (i.e. the contextual graph).

Series-parallel graphs are used in a wide range of applications, the most common being electrical circuits (Eppstein 1992) and scheduling problems (Finta et al. 1996). In these domains, flows follow all the branches simultaneously, while in a contextual graph, only one path at a time is followed because only one instantiation can be activated for a contextual element. Moreover, each total path in the graph corresponds to a unique practice developed for task realization. As presented in section 2.2.3, a contextual graph has a parallel structure for the tactical expert who wants to "keep an open mind", and has a series structure for the operational expert who prefers to focus on the details of task realization without a holistic view.

2.4. Using context in real-world applications

2.4.1. *Context and focus processing*

Enterprises develop *procedures* to address a focus in any situation but, generally, procedures result in sub-optimal solutions for any specific focus. As a result, actors develop *practices* that contextualize the procedure for addressing the specific contexts where the focus is (Brézillon 2005, 2006). In some way, they develop a *contextualized task model* in the spirit of what is shown in Figure 2.8. At the time of expert systems (see Chapter 3), the knowledge engineer elicited knowledge from experts who answered by specifying "In the specific context C1, I will apply solution S1", but the knowledge engineer only retained (problem, solution S1) whatever the context was.

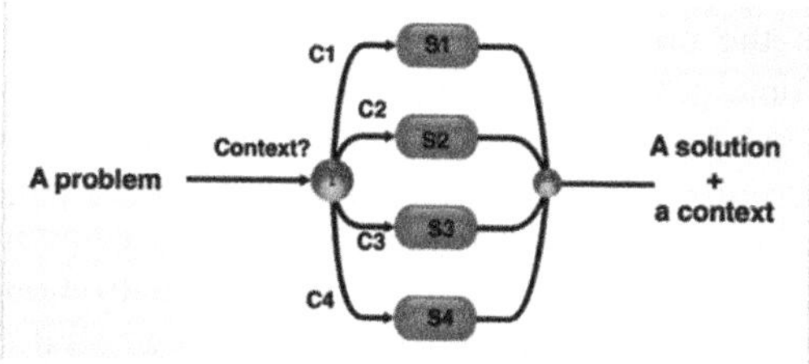

Figure 2.8. *Contextualization of a procedure by operators. For a color version of this figure, see www.iste.co.uk/brezillon/context.zip*

Formally, consider a practical unit of reasoning "IF P1, P2, … THEN C1". In a logical and theoretical reasoning (a procedure), the action leads to the conclusion C1 that is added alone to the knowledge base. In a practical reasoning (a practice), the premises stay attached to the conclusion, that is, C1 has a meaning coming from premises. A contextual element like C1 is defined in a context of use (P1, P2, etc., are true) that is lost in the procedure that aims to be applicable to a large class of similar problems, whereas a practice is a contextualized version of the procedure for the problem in a specific context. Such a practice is generally the "best practice" in that specific context, and there are as many best practices as possible contexts.

If a procedure is a representation of a task model, the modeling of practices would lead to a large number of contextualized task models because each context of the task realization is generally unique (Brézillon 2005). The contextual graphs representation formalism allows a global modeling of all of the practices developed by actors in one contextual graph. The contextual graph represents the *contextualized task model* where all the branches correspond to intertwined practices developed by actors. The practices are paths from one input to one output in the contextual graph and often a practice differs from other practices (i.e. other paths) generally on small parts, for example, by different instantiations of a contextual element (and thus different actions between the contextual node and the recombination node). Several practices use the same parts of the paths too. This contextualization of the procedure is at the tactical level, while the development of a specific practice is at the operational level (see Brézillon et al. (2007); Brézillon and Brézillon (2008), for an application concerning the self-training of car drivers).

2.4.2. *Context and actors*

A task model often comes from a decontextualization process of a large set of practices to extract what is common to the maximum number of practices. For example, in the SART project (Brézillon et al. 1997), procedures are established in a kind of CDR process by analyzing incident-solving reports (identification of the contextualization in the practices) that are formalized (i.e. decontextualized) by merging all reports on a given incident in a procedure applicable to the largest class possible of incident-solving; procedures that will be recontextualized by train drivers when needed. Procedures are exceptionally applicable directly to "real-world" incidents. In their daily work, actors have to apply the procedure in specific contexts that require adaptation of the task model (i.e. recontextualization). As a result, the actor uses knowledge and experience to interpret and decide how to (re)contextualize the procedure during their task realization in the specific context. Actors have mental models of the type of incident-solving that is generally used for this type of incident in the context at hand as operational knowledge. We observed this in the SART and TACTIC projects (Brézillon et al. 1997; Kabil et al. 2015).

An activity is described in a script that runs in a specific context. The activity is the application of an operator's strategy at the tactical level, and the practice is the description of this activity at the operational level. The notion of "activity" extends the notion of "task realization" by including the concrete implementation of a task that is realized by an actor in a specific situation with the resources at hand. The task is what an actor must do, and the activity is the (physical and mental) behavior that the actor exhibits during task realization (Sarrazin et al. 1996). Associated with each activity, there are one or more scripts that describe the ways an activity must be executed. The context determines the script to be run for an activity and concerns the result as well as the process leading to the result.

If the task realization corresponds to a personal viewpoint, the notion of activity is more adapted to a collective viewpoint. The two notions (task realization and activity) coincide if only one actor is involved. In collaborative work, several actors work together as group members. This requires an external viewpoint for "sharing out" the task realization between group members. The group activity is related to the way in which the group as a whole realizes the task. Benitez-Guerrero et al. (2012) present the development of an activity (called effective activity) as an instance of an activity model that describes the group members that can participate in the activity, how the activity can be carried out, the family of objects that can be manipulated or produced, and which roles group members and objects will play in the activity (what we call context of the activity). A group activity is more than the simple sum of actor activities because these activities are interdependent. Each group member generally intervenes at different moments during development of the group practice. Each intervention of a group member can be represented as an independent elementary task, and the interaction between members during the development of the group activity seems to be better understood as a cycle of realization of independent tasks taken in the activities of the group members. This has led us to extend the CxG formalism.

2.4.3. *Extension of the CxG formalism*

The modeling of group activity requires the extension of the CxG formalism, first for managing several actors and second, for cyclic use of a contextual graph in order to describe the development of the group activity as a sequence of independent subtasks. As mentioned previously, this supposes too the management of the shared context with the addition of particular contextual elements for modeling the management of group member interactions (turn, acceptance, etc.) during group-activity development (Brézillon 2006). The shared context becomes a crucial place for interaction management as well as task management in each actor's activity. When a subtask is realized, the shared context is analyzed at the end of the turn for determining which actor must intervene next and which independent subtask the actor must realize. We obtain a sequence of turns generated during the development of the group practice.

Because a contextual graph is directed and acyclic, its traversal corresponds to a unique subtask realization, and a simulation corresponds to a sequence of turns (i.e. successive traversals of the contextual graph). Management of the turn sequence is obtained, first, by changes made in the shared context during the previous turn, second, by the execution of actions in the subtask and, three, by the instantiation of some "particular" contextual elements defining (1) the actor (particular contextual element SENDER) who was the previous manager, (2) the actor responsible for the subtask in progress (particular contextual element MANAGER of the turn) and (3) the next manager (particular contextual elements TASK_GOAL and RECIPIENT, respectively) and its subtask concerned. The simulation of a sequence of turns requires cyclic use of the contextual graph as long as the shared context is modified during the turn. The group leadership during interaction is assured by the actor responsible for the turn, that is, for the duration of a turn (Brézillon 2017).

The modeling of the interaction between group members specifies turns and the association of each turn with the subtask of one group member. Each turn is associated with the use of the contextual graph for realizing the subtask of an actor, and the sequence of turns corresponds to the cyclic use of the contextual graph for realizing the task by the group (i.e. after several turns until the shared context stops being modified). The main point here is the modeling of the task realization in terms of *independent elementary subtasks* for each group member. Concretely, group management is represented by a general structure, called a contextual meta-graph, and a turn consists of the selection of (1) the group member that will be the manager of the turn, (2) the independent elementary subtask realized by the manager during the turn and (3) selection of the next manager and the concerned subtask. The cyclic use of the contextual meta-graph opens the possibility of backtracking into the group reasoning as well as the possibility of cycles for negotiation between two (or more) group members.

Figure 2.9 shows the organization of the contextual meta-graph in two windows, which is obtained from the CxG software. The window on the right is the graphical representation of the meta-graph with four contextual elements (blue circles, the contextual nodes with a number and the combination nodes without a number) and six activities (pink ovals containing a similar contextual subgraph with a similar structure, as shown in Figure 2.9, where each branch corresponds to an independent subtask). The window on the left presents the legends of all the items on the graphical part.

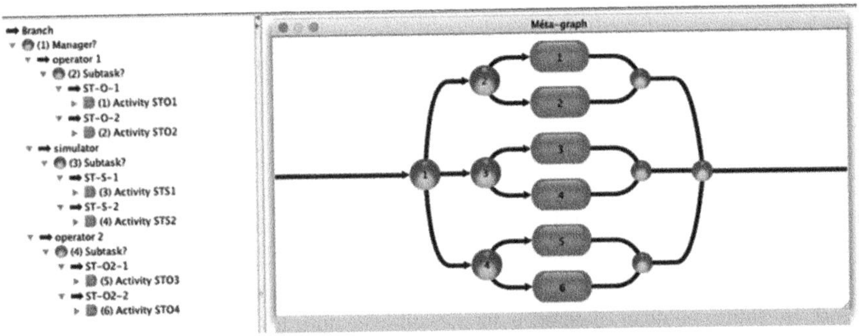

Figure 2.9. *Example of a conceptual meta-graph. For a color version of this figure, see www.iste.co.uk/brezillon/context.zip*

The explicit consideration of the shared context with the contextual meta-graph opens the door to more options than before.

– The simulation can now be stopped at the end of any turn (with the action "RECIPIENT = <nil>").

– The use of the simulation parameters introduces, in a natural way, the notion of loops for managing negotiation between actors and realizing a given subtask (or sequence of subtasks) in different contexts, thanks to the association of the contextual graph and the shared context.

– The same goal can be assigned to several actors (e.g. several reviewers for a paper submission).

– An elementary subtask may have several possible outputs depending on a contextual element instantiated in the subtask. Outputs may concern the actor or a subset of actors.

– An actor may change the instantiation of a contextual element used in the task of another actor (or several other actors).

The modeling of a group-activity development, first, is just a natural extension of the practice development by one actor and, second, the results obtained by the group-activity modeling give a new insight into modeling the task realization by an actor as an activity, with the possibility of also having a "nonlinear" representation of the actor activity, for example, checking different alternatives by simulation (cyclic use of the contextual meta-graph for one actor) before making a decision. The shared context plays a more important role than the working context because the shared context is collectively built to realize the group activity.

In this chapter, we wish to outline the value of the extension of the CxG formalism.

– A contextual meta-graph captures the accumulated experience of one or several actors that collaboratively realize a task, as said initially for one actor.

– A contextual meta-graph stays a directed acyclic graph with one input and one output, as in the initial version of the CxG formalism, but with automation and control of the cyclic use of the contextual meta-graph, thanks to the full use of the shared context.

– Traversing the meta-graph corresponds to entering in an actor's activity and realizing the specific independent subtask specified for the turn.

– Development of the group activity corresponds to a sequence of turns taken in different members' activities, a turn corresponding to the realization of an independent subtask and the management of particular contextual elements (e.g. MANAGER) controlling, on the one hand, transitions between turns and, on the other hand, the task status of the group practice development (identification of the next manager or the exit of the cycle).

– The particular contextual element MANAGER is the actor responsible for the current turn. MANAGER is instantiated for the next turn by an action during the subtask realization of SENDER.

– The particular contextual element TASK_GOAL monitors the independent subtasks to realize during the turn in the manager activity. TASK_GOAL was instantiated by an action during the subtask realization of SENDER during the previous turn.

– The shared context is composed of the contextual elements of the group activity realization and the contextual elements specific to each member's activity, information transfer between member activities and interaction management of members during group-activity development.

The introduction of the turn mechanism leads to a new type of simulation, namely the *CxG-based simulation* (Garcia and Brézillon 2015). The CxG-based simulation allows the development of complex reasoning based on the dynamical combination of a limited number of independent elementary tasks, which does not require complex modeling because the model of a group activity is built dynamically depending on the instantiation of the contextual elements at the beginning of a cycle. Moreover, the subtasks can be reused in a number of different combinations and several times with a unique representation at the implementation level. A contextual graph is similar to an *experience base* that contains a lattice of practices developed for realizing a given task. The complexity of the interactions is represented by the interplay of the particular contextual elements.

A CxG-based simulator presents a number of advantages that a model-based simulator does not. If a model-based simulation is a top-down model, a CxG-based simulation corresponds to a bottom-up model. In a model-based simulation, the whole working context is defined at the start of the simulation and stays constant during the simulation, while in a CxG-based simulation, the shared context evolves during practice development. A formal model is given initially and is compared to observations, while a practice is built progressively by the CxG simulator. In that sense, a CxG-based simulation is a particular type of simulation. The behavior of a CxG simulator is similar to the usual model-based simulator's behavior, supposing that (1) contextual elements in the contextual graph can be compared to the parameters in the formal model (a change of parameter values impacts model behavior as a change of instantiation modifies the practice developed) and (2) variables in a model-based simulation are related to the result of the progressive building of the practice corresponding to the working context.

2.5. Conclusion

This chapter sums up 25 years of research on how to model and use context in real-world applications, based on about 20 applications on a spectrum from technology-centered (SEPT, SART) to human-centered (medicine) applications, that is, from well-defined domains to more fuzzy ones. Real-world applications impose strong requirements and imply the rejection of models based on simplifying assumptions: the operator in a crisis room and the surgeon in an emergency room have to rapidly make a decision that will be definitive. Thus, the answer "Well, in 80% of the cases, the solution is…" is not an option.

On the theoretical side, McCarthy (1993) pointed out the infinite dimension of context that would appear to be a severe limitation for any modeling of context. However, we have shown that if context has an infinite dimension, all the contextual elements are not of interest for a given focus. This was made explicit by distinguishing contextual knowledge and external knowledge with respect to the focus (lesson learned 1).

Context is strongly associated with knowledge and reasoning about a real-world focus. In some sense, context has multiple aspects like knowledge as well as a process (reasoning in decision-making). Context intervenes more on the nature of the knowledge than on the knowledge itself: knowledge is contextual or not dependent on the focus. However, it is not sufficient to just say that for modeling and using context.

Context cannot be defined in an abstract way (lesson learned 2): it is dependent on a focus (task realization, problem-solving, decision-making, etc.). Contextual knowledge (the part of the context linked to the focus) is represented as contextual elements (lesson learned 3). A context-based representation formalism called contextual graphs was developed and has been implemented as a software (lesson learned 4). A contextual graph gives an inclusive representation of all the practices developed by actors for a given task realization. The nodes of a contextual graph (blue circles) correspond to contextual elements as pairs of a contextual node and a recombination node, with exclusive branches between the two nodes representing parts of the practices that actors have developed by considering different instantiations of the contextual element (i.e. in different contexts). The lesson learned 5 is that the contextual element must be differentiated from its instantiation.

The different paths from the input to the output of the contextual graph represent all the practices developed by actors for task realization. A contextual graph is a contextualized task model (lesson learned 6). The set of contextual elements in a contextual graph corresponds to the working context of the task realization, and the ordered set of contextual elements that are instantiated during the development of a practice is the proceduralized context that is used in the focus (lesson learned 7).

The extension of the CxG representation formalism combines what was done initially (i.e. modeling an actor activity) with four new "particular" contextual elements for (mainly) defining which group member is the manager, which independent elementary subtask must be realized by the manager during the turn and who will be the next manager. A group-activity development is simulated by the cyclic transversal of the contextual graph in different shared contexts (lesson learned 9). Beyond the simple simulation of the group-activity development, it is now possible to backtrack in the group reasoning to simulate alternatives, etc.

A contextual graph contains an organization of all the known practices for task realization, which makes it an experience base associated with the focus (lesson learned 8). This practice structure is organized by contextual elements that provide a coherent global picture of the group activity in various contexts (lesson learned 9).

Although the CxG formalism has been applied successfully in several fields (e.g. medicine, transportation, business management and military operations), there are some limitations. However, users have difficulty following the development of a specific practice in the contextual graph and making the decision that is associated with it. Often, experts need to quickly identify the right practice that matches the specific context at hand (especially in emergency situations), first, to evaluate a situation in order to make an immediate decision (e.g. handle an incident in the subway) and second, to infer the type of object of the reasoning corresponding to the content of a given practice. An object involved with reasoning can be either a physical object (e.g. a digital image in medicine) or a cognitive object (e.g. a diagnosis). A version of the final results as a pseudo-tree representation has been developed for this purpose to allow the focus to be quickly understood by the actors with an explicit proceduralized context and the sequence of actions to do (Garcia and Brézillon 2018) (lesson learned 10).

The revised version of the CxG formalism developed for group activity remains applicable to a unique actor with the interests of:

– nonlinear exploitation of a task realization or activity, while the task model stays "linear";

– integration of the environment state: the actor may change their mind during practice development when an unexpected event occurs (e.g. the actor selects an object and later discovers that the object is not adapted to the objective or not available);

– simulation: the actor can replay a particular subtask several times for checking different hypotheses (i.e. different contexts) before making decisions;

– exploring a decision tree: the actor can backtrack in their reasoning.

There are other potentialities to explore. First, a system using experience bases is able to explain, anticipate by simulation, incrementally complete its knowledge when failing and alert about abandoned solutions for the practice. New practices are elicited from actors when new contexts are encountered, and thus the number of practices may become very large. The capabilities of incremental knowledge acquisition, practice learning and explanation allow a system that exploits contextual graphs to become increasingly "intelligent" because it can benefit from the situation to incrementally learn new practices when it fails. In that sense, a new generation of

AI systems to be developed would be context-based intelligent assistant systems working on local experience bases corresponding to the realization of different tasks.

Second, it is possible to add into an experience base practices with failures as well as successes in the realization of the task. This could be interesting for training future actors by showing the potential danger if a contextual element is not instantiated correctly. In the SART project, the operator starting their work may replay (alone or with the previous operator) the solving of an incident to check if there is a more efficient alternative.

Third, the actors' behavior realizing the same task (in parallel or sequentially) can be studied and compared by simulation of their task realizations to determine, say, those that develop safe or risky practices (and also for statistics). This could also be used as a tool for hiring an incoming actor based on the simulation of real-world conditions.

Fourth, a contextual graph, being an experience base corresponding to the realization of one task, makes it possible to sum up the experience bases obtained from different task realizations and to compare them in order to reuse the lessons learned from one experience base in another.

2.6. References

Agabra, J., Alvarez, I., Brézillon, P. (1997). Contextual knowledge-based system: A study and design in enology. *Proceedings of the First International and Interdisciplinary Conference on Modeling and Using Context (CONTEXT-97)*, 351–362, Rio de Janeiro, Brazil.

Bau, D.Y. and Brézillon, P. (1992). Model-based diagnosis of power station control systems: The SEPT experiment. *IEEE Expert*, 36–44.

Benitez-Guerrero, E., Mezura-Godoy, C., Montané-Jimenez, L.G. (2012). Context-aware mobile collaborative systems: Conceptual modeling and case study. *Sensors*, 12, 13491–13507.

Brannen, M.Y. (2004). When Mickey loses face: Recontextualization, semantic fit, and the semiotics of foreignness. *Academy of Management Review*, 29, 4, 593–616.

Brézillon, P. (2005). Task-realization models in contextual graphs. In *Modeling and Using Context (CONTEXT-05)*, Dey, A., Kokinov, B., Leake, D., Turner, R. (eds). Springer Verlag, Berlin.

Brézillon, P. (2006). Some characteristics of context. In *Advances in Applied Artificial Intelligence*, Ali, M. and Dapoigny, R. (eds). Springer Verlag, Berlin.

Brézillon, P. (2017). Contextual modeling of group activity. In *Modeling and Using Context*, Brézillon, P., Turner, R., Penco, C. (eds). Springer Verlag, Berlin.

Brézillon, P. (2018). CxG-based simulation of group activity [Online]. Available at: https://www.openscience.fr/CxG-based-simulation-of-Group-Activity.

Brézillon, P. and Brézillon, J. (2008). Contextualized task modeling. *Revue d'Intelligence Artificielle*, 22(5), 531–548.

Brézillon, P. and Pomerol, J.-C. (1999). Contextual knowledge sharing and cooperation in intelligent assistant systems. *Le Travail Humain*, 62(3), 223–246.

Brézillon, P. and Pomerol, J.-C. (2001). Is context a kind of collective tacit knowledge? [Online]. Available at: http://citeseerx.ist.psu.edu/viewdoc/download?doi=10.1.1.16.991&rep=rep1&type=pdf.

Brézillon, P., Gentile, C., Secron, M. (1997). SART: A system for supporting operators with contextual knowledge. *First International and Interdisciplinary Conference on Modeling and Using Context (CONTEXT-97)*, 209–222, Rio de Janeiro, Brazil.

Brézillon, P., Pasquier, L., Pomerol, J.-C. (2000). Representing operational knowledge by contextual graphs. In *Advances in Artificial Intelligence*, Monard, M.C. and Sichman, J.S. (eds). Springer, Berlin, Heidelberg.

Brézillon, P., Drai-Zerbib, V., Baccino, T., Therouanne, T. (2006). Modeling collaborative construction of an answer by contextual graphs. *Proceedings of the IPMU*, 11–13 May, Paris, France.

Brézillon, J., Brézillon, P., Tijus, C. (2007). Improving driver's situation awareness [Online]. Available at: https://www.researchgate.net/publication/282062789_Improving_driver's_situation_awareness.

Brézillon, P., Attieh, E., Capron, F. (2014). Modeling glocal search in a decision-making process [Online]. Available at: https://www.researchgate.net/publication/262912538_Modeling_Glocal_Search_in_a_Decision-Making_Process.

Clancey, W.J. (2002). Simulating activities: Relating motives, deliberation, and attentive coordination. *Cognitive System Research*, 3(3), 471–499.

Eppstein, D. (1992). Parallel recognition of series-parallel graphs. *Information and Computation*, 98(1), 41–55.

Fan, X., Zhang, R., Li, L., Brézillon, P. (2011). A context-based framework for improving decision making in scientific workflow. *Proceedings of the 3rd International Conference on Computer Research and Development (ICCRD-2011)*, Shanghai, China.

Finta, L., Liu, Z., Mills, I., Bampis, E. (1996). Scheduling UET-UCT series-parallel graphs on two processors. *Theoretical Computer Science*, 162(2), 323–346.

Furnas, G.W. (1986). Generalized fisheye views [Online]. Available at: https://citeseerx.ist.psu.edu/viewdoc/download?doi=10.1.1.124.1233&rep=rep1&type=pdf.

Garcia, K. and Brézillon, P. (2015). A contextual model of turns for group work. In *Modeling and Using Context (CONTEXT-15)*, Christiansen, H., Stojanovic, I., Papadopoulos G.A. (eds). Springer, Berlin, Heidelberg.

Garcia, K. and Brézillon, P. (2018). Model visualization: Combining context-based graph and tree representations. *Expert Systems With Applications*, 99(1), 103–114.

Grimshaw, D.J., Mott, P.L., Roberts, S.A. (2000). The role of context in decision making: Some implications for database design. *European Journal of Information Systems*, 6(2), 122–128.

Kabil, A., Brézillon, P., Kubickki, S. (2015). Contextual interface for operator-simulator interaction. In *Modeling and Using Context (CONTEXT-15)*, Christiansen, H., Stojanovic, I., Papadopoulos, G.A. (eds). Springer, Berlin, Heidelberg.

Karsenty, L. and Brézillon, P. (1995). Cooperative problem solving and explanation. *International Journal of Expert Systems with Applications*, 4, 445–462.

Landauer, C. and Bellman, K. (1993). The role of self-referential logics in a software architecture using wrappings. *Proceedings of the 3rd Irvine Software Symposium*, Irvine, CA, USA.

McCarthy, J. (1993). Notes on formalizing context. *Proceedings of the 13th IJCAI*, 555–560.

Miller, G.A. (1956). The magic number seven plus or minus two: Some limits on our capacity for processing information. *Psychological Review*, 63, 91–97.

Pomerol, J.-C. and Brézillon, P. (1999). Dynamics between contextual knowledge and proceduralized context. *Proceedings of the 2nd International and Interdisciplinary Conference on Modeling and Using Context*, Trento, Italy.

Pook, S., Lecolinet, E., Vaysseix, G., Barillot, E. (2000). Context and interaction in zoomable user interfaces. *Proceedings of the Working Conference on Advanced Visual Interfaces*, Palermo, Italy.

Sarrazin, P., Biddle, S., Famose, J.-P., Cury, F., Fox, K., Durand, M. (1996). Goal orientations and conceptions of the nature of sport ability in children: A social cognitive approach. *British Journal of Social Psychology*, 35, 399–414.

Schank, R.C. (1982). *Dynamic Memory, A Theory of Learning in Computers and People.* Cambridge University Press, Cambridge.

Sowa, J.F. (2000). *Knowledge Representation: Logical, Philosophical, and Computational Foundations*. Brooks Cole Publishing Co, Pacific Grove, CA, USA.

3

Toward Pragmatic Context-Based Intelligent Systems

Context-based intelligent systems are those that not only focus on problems but also on the context in which the problems are to be solved. Instead of simply applying general (or even knowledge-based) problem-solving methods, they *contextualize* the methods in order to tailor them to the situation at hand. To do this, they rely on contextual knowledge – knowledge about how to behave in a situation – as well as knowledge about the context, in order to assess the current situation and bring the correct contextual knowledge into focus. In general, there have been two approaches to context-based systems (indeed, to everything) in AI: formal and pragmatic, with the former focusing first on the formal properties of and relationships between contexts and only then attempting to solve real problems, while the latter focuses first on solving real-world problems and then on abstracting general principles from what has been done. We argue here that the pragmatic approach is most useful for building systems to solve real-world problems. We discuss the requirements needed for pragmatic context-based systems in general, as well as for two broad classes, in particular: context-based intelligent assistant systems (CIASs) and context-based intelligent autonomous agents (CIAAs). We also draw on our own work in pragmatic context-based intelligent systems, for example.

3.1. Introduction

For much of its early history, there was a healthy tension in the field of artificial intelligence (AI)[1] between, on the one hand, formal theories (e.g. theorem proving

Chapter written by Roy M. TURNER and Patrick BRÉZILLON.

1 AI refers here to the entire field, not just its machine learning subfield (which itself includes deep learning/neural networks). As AAAI (Association for the Advancement of Artificial

rooted in formal logic) and system engineering and, on the other hand, appeals to "softer" theories arising from the emerging field of cognitive science (e.g. means–ends analysis) and pragmatic, exploratory approaches. These are often summarized as the "neat" versus "scruffy" approaches to AI. Research on the neat side tried to systematize knowledge and reasoning in a top-down manner, starting with theory and working toward creating AI systems embodying those theoretical frameworks. The "neat" approach is rooted in the logicism that refers to the school of thought, founded by McCarthy, that intelligent reasoning can be expressed within formal logic. Research on the scruffy side was more pragmatic and recognized the fledgling nature of AI when compared to the immense complexity of real-world problems; it instead focused first on finding solutions, often guided by cognitive science, to those problems, only later abstracting what was learned to propose (often informal) theories. Though there were often spirited disagreements between the practitioners of each style of AI, there was also a healthy bidirectional flow of ideas that improved both, with theory contributing rigor and structure to the more problem-based, bottom-up scruffy approaches, while the latter pointed out shortcomings and gaps in theoretical approaches and suggested solutions.

By the 1980s, however, the balance in the field had shifted toward formal, engineered systems, with knowledge and reasoning methods largely based on theorem proving, rule-based expert systems, Bayesian networks and fuzzy logic. Unfortunately, as recognized by the scruffy side of AI, formal/mathematical systems of this sort can only go so far. They face severe limits in the real world, which is extraordinarily complex and messy, with missing information, uncertainty and imprecision, and with the presence of other, possibly unknown and unpredictable, agents and processes acting to change the world in unexpected ways. It is all but impossible to create general, formal models of the world and reasoning mechanisms that can guide an agent's or application's behavior, so that it is appropriate for all the situations it may encounter in the complex, highly variable real world.

We argue here for a strengthening and renaissance in AI's pragmatic roots, especially as it pertains to solving real-world problems. We further argue that a promising route to doing this is to embrace and build on the pragmatic side of research on context-based systems.

Research in the past 25 years on modeling and using context has shown that one way to successfully attack complex worlds and problems is to transform knowledge about how to operate in a domain – formal and theoretical knowledge, as well as heuristic and even tacit knowledge – into *operational*, contextualized knowledge that informs an agent or application about how to behave in the particular context it

---

Intelligence) states, AI is the "scientific understanding of the mechanisms underlying thought and intelligent behavior and their embodiment in machines" (from aaai.org).

faces in a particular situation. This *contextualization* process adapts the agent's knowledge so that it is specific to the task at hand, the current situation (including the local environment, resources that may or may not be present in that situation, any users' preferences, etc.) and the system's and user's past experience. It takes the intractable problem of knowing how to behave in any situation and turns it into a set of more tractable problems: how to behave in particular contexts. To do this successfully requires attention to context as an object of study in its own right.

In the remainder of this chapter, we first look at the evolution of AI in more detail, focusing specifically on the formal/pragmatic balance. Next, based on lessons learned over the past 25 years about context research (see Chapter 1), we discuss what is necessary to create pragmatic context-based intelligent systems in general, as well as for two important kinds, specifically: context-based intelligent assistant systems (CIASs) and context-based intelligent autonomous agents (CIAA).

3.2. Evolution of AI systems

3.2.1. *Formal versus pragmatic acontextual approaches*

Early AI systems were general-purpose systems, such as the General Problem Solver (GPS; Newell and Simon 1963), theorem provers (Davis and Putnam 1960), early planning systems (e.g. STRIPS; Fikes and Nilsson 1971) and game-playing programs centered around the minimax algorithm. These are so-called "weak" methods, referring to their lack of focus on modeling and incorporating knowledge, specifically domain-specific knowledge. Weak methods enjoyed great success across a wide range of problem types (creating plans, playing games such as chess, proving mathematical theorems, doing symbolic integration and even performing intelligence tests). However, they were largely limited to highly formal problems or problems in toy worlds, since they were generally unable to handle uncertainty, incomplete knowledge or dynamic situations. This was partly responsible for the first "AI winter", when funding for AI largely dried up.

The mid-1970s saw the emergence of the first so-called "strong", or knowledge-intensive, methods. These were *expert systems*, the name referring both to the incorporation of expert knowledge in them and their ability to perform at or above the level of human experts in some real-world domain. DENDRAL (Lindsay et al. 1980), for example, often considered the forerunner of expert systems, in the mid-1960s determined the structure of organic molecules from mass spectra and other data using domain knowledge obtained from a Nobel prize-winning chemist. Researchers who worked on DENDRAL went on to create the first rule-based expert system, MYCIN (Shortliffe 1974). MYCIN was able to diagnose bacteremia (blood-borne infections) better than almost all human diagnosticians. It gave rise to an entire family of expert systems, as well as the field of knowledge engineering. Its associated projects pioneered work on meta-reasoning, that is, reasoning about reasoning, useful for both knowledge acquisition and for explanation of the system's reasoning to humans.

Expert systems were a major reason for the end of the AI winter (e.g. Toosi et al. 2021), since they promised the ability to deal with problems in highly uncertain, real-world domains, where there was often incomplete or conflicting knowledge. These were *pragmatic* systems, less concerned with the formal properties of solving problems than with *actually* solving them. The expert knowledge on which they depended was *heuristic*, including heuristic estimates of the certainty of conclusions based on facts, and their problem-solving method was also heuristic with respect to how uncertainty was managed and propagated, and how rules were selected to be applied (for forward-chaining systems).

Expert systems' knowledge was almost always "flat", that is, non-structured, consisting of a set of independent rules. This allowed the knowledge engineer to implement updates easily and also one of the great strengths of expert systems, a clean separation of knowledge (the rules) and processing (the "inference engine" that processed the rules). However, this also meant that any context-specific operational knowledge was buried in the rules themselves, and it was up to the knowledge engineer and domain experts to make sure that the rule antecedents allowed such knowledge to be brought into focus in the right context.

3.2.2. *Formal consideration of context*

Formal AI researchers were, by the mid-1990s, beginning to recognize that acontextual approaches were not able to handle the complexity and context-specificity of knowledge involved in real-world problems. While a primitive form of context representation, the situation calculus (McCarthy and Hayes 1969) was developed to allow atemporal theorem provers to reason about facts holding in different situations, this was insufficient to truly represent and reason about context. McCarthy (1993) was one of the first to recognize that it is useful to identify contexts as objects about which to reason directly. He introduced the *ist()* predicate to explicitly state that something is true in a particular context, and also developed "lifting" axioms to assert that if some proposition is true in one context, another proposition is then true in another context. Similar functionality was developed by Giunchiglia (1993) and others in their work on "bridge" rules.

Even though this was a good start, it suffered from the treatment of contexts as discrete, unstructured objects relative to which propositions could be assumed to be true. The focus was on what to do in a context and how to move from one to another without really focusing on what a context *is*.

3.2.3. *Pragmatic consideration of context*

Expert systems were the first pragmatic approach to have some recognition of the need for contexts, although in a very simple form. As previously mentioned, contextual knowledge *was* present in even the first expert systems, but it was implicit in the experts' knowledge and rules had to be crafted to ensure that the correct knowledge was brought into focus in the correct situation. This led to a plethora of repeated context-related antecedent clauses across the rule base, since many rules could apply in a given context.

A simple mechanism to avoid this, similar to introducing a situation parameter to predicates in the situation calculus, was to have a fact in the systems' working memory that explicitly stated what the context was. Rules could then just refer to that in their antecedents to ensure that the right rules are triggered and fired in the right context. Other rules, usually triggered only when there was nothing left to do in a given context, were then added to transition from context to context. This approach suffered from the same problems as noted for formal approaches above.

A more structured approach to context was taken in Centaur (Aikins 1980), where prototypes (i.e. frames; Minsky 1975) were used to represent not only prototypical cases and problem-solving strategies for medical diagnosis, but also the contexts in which problem-solving knowledge should be used. Other, similar work was done in the MDX (Chandrasekaran et al. 1979) and MDX-2 (Sticklen 1987) projects. Here, rules were organized into packets organized in a hierarchy mirroring domain knowledge. For example, a rule packet may contain operational knowledge about how to diagnose a category of disease, such as cardiopulmonary disease, as well as links to other packets of knowledge for specializations of the disease (e.g. cardiac disease, pulmonary disease) or for related diseases.

A similar idea was developed in the SEPT project (Brézillon 1989), a system for assisting operators in French EHV (Extra High Voltage) substations. Here, a standard expert system was found to be far from optimal, since the system had to diagnose different kinds of equipment, and the diagnosis of one kind (e.g. a problem related to a circuit breaker) is radically different from the diagnosis of another (e.g. a protective relay). The system evolved from using a flat rule base of about 600 rules the inference engine tried to instantiate (with about 20 possible values for each rule variable), to the one using rule packets to only focus on those rules germane to the kind of equipment being diagnosed. This reduced diagnosis time by a factor of 50, which was important in this domain, where diagnoses needed to be made within two seconds. Rule packets were selected based on rules such as:

```
IF Equipment(X) = circuit_breaker THEN
   call rule_packet(circuit_breaker,X).
```

A second generation of expert systems, *knowledge-based systems* (KBSs), were model-based rather than rule-based (Brézillon and Pomerol 1996; Smith 1985). Although this introduced some much-needed structure to the systems' knowledge and allowed some benefits in both knowledge acquisition and use, they largely forgot the lessons learned about context from the first generation. This was unfortunate, since organizing knowledge by context instead of or in addition to the structure of the domain can dramatically simplify knowledge acquisition and ensure its proper use.

The usefulness of expert systems and KBSs depended on who used them and in what context. They tended to rely on a particular style of reasoning (production systems) that tended to be far removed from how experts actually reason. This prompted experts and knowledge engineers to add new rules to control the firing of existing rules, in order to achieve the same kinds of results as a human. Meta-reasoning was needed to explain to the user how the system arrived at its conclusions in a manner they could understand. Although there were many successes (see, for example, Bobrow and Brady (1998)), the systems rapidly became intractable.

In addition, just the introduction of expert systems/KBSs in an organization often ran into problems. This was for several reasons, including power-sharing changes brought on by the system, user lack of acceptance of an inscrutable oracle, the (justified) fear of being replaced, the fear of being spied upon by the system and all of the other (rational or irrational) fears faced when introducing any automation into an organization, but amplified by the capabilities and intended use of the systems.

Some very valuable lessons have been learned from expert systems and KBSs, however, that can inform future pragmatic systems. First, unexpected situations are the norm rather than the exception. The systems by their nature were not designed to solve the unexpected, and this forced the users to face difficult problems on their own just when they could most use the help. Second, expert knowledge rapidly loses its highly contextual nature when acquired and coded in a system, unless extraordinary care is taken by the knowledge engineer to adjust the antecedents of rules to match the knowledge's context of use. An expert, when asked if a fact or procedure is correct, will often respond with "it depends on the context". To them, there is a mapping from (problem, context) to solution. However, very often in expert systems/KBSs, the mapping is simply from problem to solution, with little influence from the context of knowledge application. Third, the systems necessarily have limited and incomplete expert knowledge, which limits their ability to solve problems. This contributes to the occurrence of exceptions, since situations that are routine for the human expert may become exceptional for the system if the correct pieces of knowledge were not acquired. Fourth, the systems are often given the role of primary problem-solver, with users relegated to being little more than simple data gatherers. Fifth, the systems tend to be *brittle*, that is, their performance does not degrade gracefully at the edge of or beyond their area of knowledge, and indeed, they tend not to know where that boundary is. Coupled with the usual role of system-as-oracle, this can lead to disastrous consequences when the system fails outright or, worse, provides the wrong answer with inappropriately high confidence. While this problem arises due to holes or errors in their domain knowledge, a lack of general-purpose reasoning mechanisms and knowledge, and the relegation of the user to a support role, a major cause is the lack of attention to the context in which particular knowledge is applicable. And, sixth, the systems were usually targeted to solve the 80% or so of relatively simple problems that would be encountered, but in most situations, the user is experienced enough in their field to handle some problem-solving and is more in need of support solving the 20% or so of hard problems.

One line of research leading toward context-based reasoning evolved directly from the experiences of the KBS community: *joint cognitive systems* (JCSs; Woods 1985). Here, the focus was on the complementary nature of competence and skills of the human and the system with respect to task realization, in particular to exploit their cognitive asymmetry.

In a JCS, the two work as partners in problem-solving. The JCS approach requires the system to have a simulation component. The system uses this and its knowledge to decide what action should be taken at the moment, recognize the path the user is on and the next action they are about to take, and alert the user of risky consequences. The user, however, remains the decision-maker.

This moves the focus from modeling the task to modeling task *realization*. This has to take into account the context, and it results in different *practices* being developed by the user to realize the task in different contexts.

Unfortunately, developing a complete, competent model of effective practice development was not possible in JCSs, since the context was not made explicit, and there are as many practices as there are contexts.

Brézillon and Abu-Hakima (1995) point out that the line of reasoning in a JCS developed during problem-solving is a co-construction of the system and the user, and that explanations become an intrinsic part of the construction. Each of the agents, human and computer, may modify the joint line of reasoning at any time, which will cause a modification in the private line of reasoning of the other agent. Thus, the context during problem-solving includes direct information from the agents, mainly based on their interactions and on the user's actions on the real-world process. This means that the AI system needs to understand the user(s) through their decisions, actions and behaviors, not from some abstract model or profile from a library. Thus, it would be better to think of this kind of system as an *intelligent assistant system* (IAS) rather than a JCS, since the goal is to help the users with their tasks, not replace them.

AI systems evolved from focusing solely on the task and the world (formal systems, expert systems/KBSs), to the system and user working in complementary ways (JCSs), to a fusion of system and user for reasoning about task realization (IASs).

The idea of intelligent assistant systems is not new, of course, and has long been a staple of science fiction. In Pohl's novel *The Age of Pussyfoot* (Pohl 1969), for example, each person needed a mobile device called a "Joymaker" to buy products, communicate with others, watch TV, monitor the children and even receive medication. In the story, "computer-centered" humans could not live in society without their Joymakers. This type of computing is of course now everywhere: mobile phones, smart houses, connected objects (Internet of Things), etc.

Most IASs strive to understand the current context, including the task at hand, any real-world processes involved, and the user's goals and intentions as reflected by their behavior. An IAS may learn a practice by observing the behaviors of an actor in relation to the task model and knowledge of the world: by understanding the evolving context.

Unfortunately, IASs' knowledge of the context is often sparse. It may consist of location (GPS), Internet-based services (weather, etc.) and expected properties of the task, as well as a user profile based on statistical considerations and augmented or selected with the help of available sensors (e.g. biometric sensors for identification or determining the physiological state). However, most IAS lack any attention to the full panoply of facts, knowledge and detail that is the total context in which the IAS and its user operate.

Apart from the line of research leading from expert systems to IASs, other pragmatic research in AI also addressed the need to take context into account. SOAR (Laird et al. 1987) was proposed as an architecture for general intelligence based on production rule-like operators and a reasoning mechanism designed to organize their application to realize intelligent actions, including complex activities such as hierarchical planning. SOAR (now just "Soar") is based on the *physical symbol system hypothesis* (Newell and Simon 1976): that a set of symbols and a means to manipulate them is the necessary and sufficient condition to support general intelligence.

While Soar does not explicitly represent contexts as first-class objects, it does create and reason within them dynamically as a result of its major feature, subgoaling, which creates new *problem spaces* – its analog to contexts. When problem-solving stalls in a given problem space, the system creates a new problem space with the goal of resolving whatever impasse caused it to stall in the old one. The subgoaling process is recursive. When an impasse is solved, Soar uses its "chunking" learning mechanism to compile the results of problem-solving done in a problem space into a new rule. The learned rule associates the significant parts of the situation before the impasse with the result. The next time a similar situation is encountered and an impasse potentially exists, instead of subgoaling, the new rule will be used to avoid the impasse.

Unfortunately, although Soar treats problem-solving spaces as contexts, it does not make use of contextual knowledge in any real sense; for example, the context in which an impasse occurs does not impact how it is solved. In addition, once an impasse is solved, all information about how it was solved in that context is abandoned.

Other earlier work from the cognitive side of AI also came to realize the central role played by context in effective, real-world reasoning. Cullingford (1981) and Schank and Abelson (1977) developed the idea of *scripts* as a knowledge structure that could provide the context necessary to not only interpret utterances in a story, but also fill in missing pieces of knowledge – even those that would be unreachable by inference alone due to being conventional rather than goal-based. For example, a restaurant script would allow the two-sentence story:

> John went to a restaurant.
>
> He got a hamburger.

to be understood to the extent that later questions posed to the system could be answered with minimal reasoning: Did he pay for the hamburger? Did he eat the hamburger? Both would be answered affirmatively based on the knowledge of the context supplied by the script.

A generalization and extension of scripts, *memory organization packets* (MOPS; Schank 1983) were later proposed as a way to organize memory in a human-like manner; these could also provide contextual predictions. Unlike the discrete contexts found in formal systems, scripts and (especially) MOPs accounted for how a context can continuously evolve.

Case-based reasoning (CBR; see, for example, Kolodner (1987)), which came out of the same line of research, can be seen as one approach to both representing contexts and operationalizing problem-solving knowledge by recording practices. Upon completion of a problem-solving episode, a record of it is stored as a case in a memory (often based on MOPs). When another problem situation "reminds" the system of the case, it is retrieved and used, either to suggest solutions directly, provide (operationalized) inferences and procedures to reuse, or both.

A problem with CBR, however, is that it is too granular and specific. Its "contexts" are individual situations, and the system has to reason about how to apply the knowledge in them each time they are retrieved from memory. Moreover, CBR conflates "situation" with "context", when the two are in fact distinct. For example, we may view the current situation as being something that is goal- or problem-based – for example, regulating subway traffic, or diagnosing a patient's problem – that occurs *in* a context – for example, rush hour, or in the setting of a hospital.

Regardless, CBR has been a very successful approach spanning a wide range of applications and even reasoning styles, and it led to one of the strains of context-based reasoning approaches discussed later.

3.3. Pragmatic context-based intelligent systems

Context is so important in *any* form of reasoning that it should be taken into account as a primary determinant of how to behave. Indeed, it would be hard to imagine any system that was not sensitive and conformant to its context as being intelligent in any real sense.

While many of the approaches discussed above paid *some* attention to context, with some even attempting at least rudimentary representations of it, none of them could be realistically called *context-based* intelligent systems: that is, intelligent systems that consider context and contextual knowledge as an integral part of doing their job.

In this section, we examine what is needed to create context-based intelligent systems that can solve real-world problems: pragmatic context-based intelligent systems. The first few sections focus on the required attributes and capabilities of any pragmatic context-based intelligent system. In the final two sections, we look at two important classes of such systems, those that are intelligent assistants or joint problem-solving systems and those that are primarily autonomous.

3.3.1. *Explicit context representation*

Experience with the systems discussed above brought the realization that context is important in AI systems. This led to attention being focused on context itself, and how to represent and use it. In this section, we outline some of the first attempts in this regard. For a more complete discussion of the work done by the community of researchers focused directly on context and its modeling and use, we refer the reader to Chapter 1, where we pay particular attention to the contributions from the community centered around the CONTEXT[2] biennial conference series.

One of the first insights from research on context as a subject of study in its own right was that a system must represent it explicitly in order to use it effectively. This was one of the problems with prior attempts to use contextual reasoning. Without an explicit representation of a context to serve as the organizing locus for knowledge about it, that knowledge is scattered across the agent's problem-solving knowledge, making it difficult to acquire the knowledge in the first place, and update it and keep it consistent thereafter. Human experts naturally organize their knowledge with respect to contexts and can often delineate features of contexts and useful facts about them; this should be capitalized on to acquire knowledge efficiently. Similarly, when the system learns something new about a context, for example, a new practice that better achieves a goal in it, then it is important that it be able to efficiently find all of the relevant knowledge so that it can be updated.

Explicit representation helps the system to assess its current context, in much the same way that explicit knowledge about diseases helps a doctor diagnose the particular disease present in the current patient. Different contexts can be examined to compare their features and predictions with the current situation to determine which fit best and thus explain the situation. Knowledge about contexts can also help the system realize when the context has changed and what the new context is. Knowledge associated with a context representation can also help predict future contexts that may occur (see Chapter 7).

Another insight from context research is that a context-based system must have a means to represent the richness of contexts and contextual knowledge, not just the kind of simplistic representations discussed above. However, representing context explicitly is tricky. McCarthy (1993) formally showed that context cannot be described completely because, on the one hand, context is strongly related to the knowledge and reasoning of the domain and, on the other hand, it has an infinite dimension, not least because contexts can be embedded in other contexts to an arbitrary level of nesting. It is difficult, if not impossible in principle to know which

2 International and Interdisciplinary Conference on Modeling and Using Context.

of the potentially infinite facts about the world comprises part of a given context, a variant of the *frame problem* (Shoham 1987). Thus, it is difficult to define context as an abstract object, even with discrete contexts.

To avoid facing this problem, many recent context-related research communities (context-aware applications, ubiquitous computing, pervasive computing, context-driven testing, etc.) in different disciplines use a pragmatic, bottom-up approach, in which contextual elements of interest are directly drawn from the actor, the task, the situation and the environment. This is in the right direction, but unfortunately, the work in these communities often only deals with an impoverished idea of what the context is, focusing only on a few simple features, such as the user's identity, the location of use, etc. In addition, context is often not viewed as a first-class object, which severely limits reasoning about the context itself.

Several researchers in the larger context research community have attempted to address this problem in a more complete fashion. Two lines of research by the authors, for example, are illustrative of how a richer representation of context can be made explicit and used by an agent.

Brézillon et al. (2000) proposed a conceptual framework for context modeling with an implementation called *contextual graphs* (CxG). This approach relies on the operational definition of context as what constrains a system's problem-solving focus without directly intervening in it (Brézillon and Pomerol 1999). This approach was developed in the domain of handling problems arising in a subway system. "Focus" here refers to things such as task realization, decision-making, problem-solving, incident solving, diagnosis or subway platform configuration. The definition implies that while context is related to problem-solving knowledge, it includes (at least) knowledge of the domain, the task, the situation and the user.

At any given moment, the system's knowledge is composed of contextual knowledge that is related to the current focus and other knowledge – external knowledge – that is not. A subset of contextual knowledge is selected, organized, assembled and structured to participate in the focus. This is the *proceduralized context* used in the current step of problem-solving.

Turner (1998, 2014) also proposed a framework, *context-mediated behavior* (CMB), for representing context as a rich, first-class object. This work is part of a reasoning mechanism called *schema-based reasoning* (Turner 1994), which grew out of and represents a generalization of case-based reasoning. In CMB, a distinction is made between "situation" and "context". Here, an agent's *situation* is comprised of all aspects of the agent, other agents, the task and the environment, akin to the idea of "state" in work on research. It is this that has an infinite dimension. In contrast, a *context* is considered to be a class of situations that have some common, important implications for how the agent should behave. It is contexts, not situations, that are represented explicitly.

In this approach, known contexts are represented as contextual schemas (*c-schemas*) that contain descriptive knowledge about the class of situations (e.g. what entities are expected, their features, what concepts mean in the context, etc.), as well as prescriptive knowledge of how to behave (e.g. how to handle various unanticipated events that may occur, what goals to prioritize/de-prioritize, how best to achieve goals in the context, behavior parameter settings, etc.).

CMB envisions a context manager that runs in the "background" and constantly strives to assess the agent's current context by diagnosing it as one or more known contexts. The "foreground" part of the agent (i.e. the performance element, in the parlance of machine learning) then uses knowledge from the current context object (a composite of matching c-schemas) to carry out its tasks in a context-appropriate way. This object can be viewed as analogous in a way to CxG's proceduralized context.

In both of these approaches, CxG and CMB, accomplishing a task involves building an explicit context-specific model at the time it is needed, for example, during practice development or reuse. This approach is close to Clancey's (2002) view that solving a particular problem (e.g. diagnosing a patient) involves creating incrementally situation-specific models, where "situation-specific" refers to a particular case, setting or scenario.

With such context-specific models of problem-solving, tasks are realized in context, using operationalized procedures and other problem-solving knowledge. In other words, the context models provide the background against and within which practices are developed.

A problem faced by such context-based systems is how to transform domain knowledge and general, perhaps theoretical and formal, procedures for operating in the domain into operational knowledge about how to solve problems in particular contexts faced by the system. This requires contextualization of the general knowledge to adapt it to the situation, the local environment, the system's past experiences, the task at hand and, for those systems with users, their past experiences and preferences. It requires that context becomes an object of study in its own right.

Given the need for context-associated operationalization of knowledge, in context-based systems, knowledge bases really become context-based knowledge bases – that is, "experience bases" the system can use to solve problems in highly situation-specific ways. In CxG, for example, the system's experience base allows an intelligent assistant to reason about and help a user during task realization by modeling their behavior (i.e. reasoning in terms of practices, not just procedures). In CMB, the experience base – the c-schema repertoire – provides the building blocks for context assessment and the knowledge of which procedure variants (practices) and other knowledge can be used to allow the agent to behave appropriately in the situation.

3.3.2. *Context assessment mechanism*

All systems also need the ability to recognize the context they are in, although different systems may have different ways of doing this. This could range from very simple feature-based pattern-matching (of features of the world to expected features of context), to more complex theorem proving-like mechanisms, to very complex diagnostic approaches.

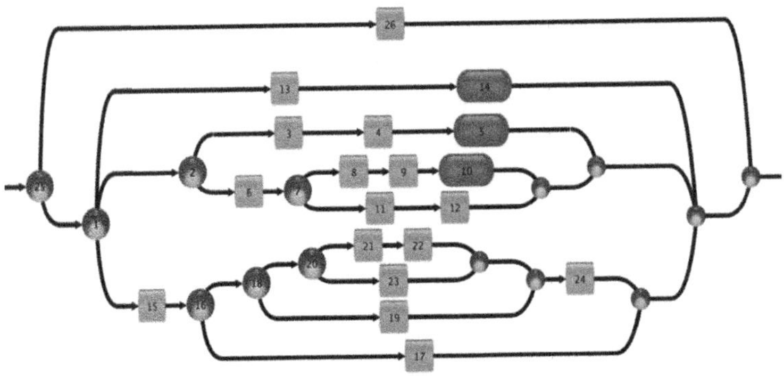

Figure 3.1. *A contextual graph. The circles are variables, the instantiation of which determines the branch followed. For a color version of this figure, see www.iste.co.uk/brezillon/context.zip*

One approach to this is exemplified in the contextual graph (CxG) framework. In part due to its focus being the operational level of problem-solving (rather than tactical or strategic), CxG takes a granular, continuous approach to modeling the context. Figure 3.1 shows a contextual graph. At no point does CxG commit to being in a particular known context, but rather assumes that the context is continually changing during problem-solving: the evolving contextual graph *is* the context model, and the system (or the system and its user) are in the context specified by the particular instantiation of the variables that led to the current point in problem-solving (Brézillon and Gonzalez 2006).

Context assessment is then an ongoing process of determining the values of these variables, in order to determine which branch of the graph to follow (or that is being followed, if observing the process). This is done by the system as it monitors the user's actions and the state of the world. With each instantiation, the context assessment changes.

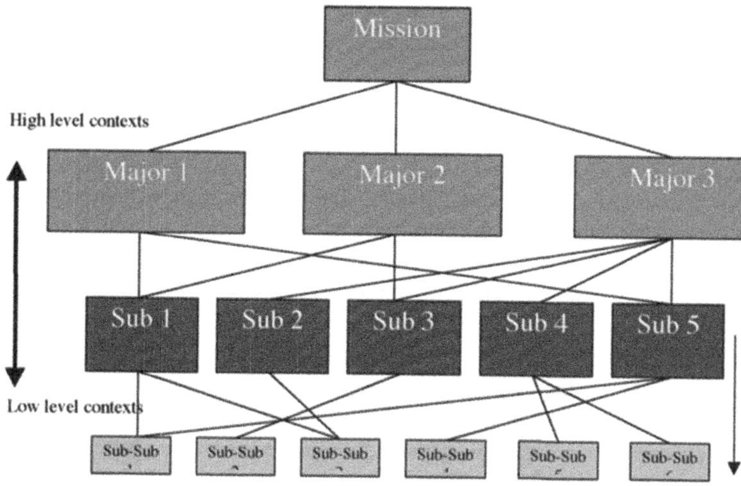

Figure 3.2. *A mission and its contexts in CxBR (from Brézillon and Gonzalez 2006). For a color version of this figure, see www.iste.co.uk/brezillon/context.zip*

Another approach is that of the context-based reasoning (CxBR) framework. Here, contexts are discrete, rather than continuous as in CxG. Figure 3.2 shows the organization of representations for possible contexts for a mission. The system monitors the current situation to determine what the context is. In some implementations of CxBR, the contexts themselves contain knowledge about which other contexts are appropriate in the situation, while in at least one other, a neural network was used to find and assess the context (Gonzalez et al. 2006).

A third approach is that of the context-mediated behavior (CMB) framework. In CMB, context assessment is done by using a diagnostic process (see Chapter 5) based on medical diagnosis (Miller et al. 1982). Candidate context representations (contextual schemas) are identified based on superficial matches to the current situation's features; these "evoked" c-schemas are then grouped into sets based on explaining the same set of features, and then information is sought to separate one context hypothesis in each set from the others enough to be considered part of the context assessment. The leading hypotheses from all sets are then merged to form the assessment of the current context.

3.3.3. *Context transitioning mechanism*

Since context is dynamic, all systems also need a way to transition from one context to another. As we saw above, some approaches use a very small-grained view of the differences between contexts and move from context to context accordingly, so that there is very smooth, almost continuous movement from one to another. CxG, for example, changes its model of the context with each variable instantiation in the current contextual graph.

Other approaches have discrete contexts, and will thus transition in discrete, abrupt steps. CxBR exemplifies this. Here, knowledge about how to transition is represented as *Sentinel Rules* that monitor the environment and change the context accordingly (Brézillon and Gonzalez 2006). Changes from major to sub-contexts (or sub-sub-contexts) are guided by the hierarchy of contexts.

Still other approaches occupy the middle ground between fine- and coarse-grained context transitions. For example, CMB's context manager module monitors the situation in comparison to the set of context representations currently in use to model the context. As the situation changes, the context assessment can change abruptly from one entire context to another, but it is more likely that only some of the component context representations will be replaced in the overall representation. For example, if an autonomous underwater vehicle is performing a search mission in a harbor with low power, when it exits the harbor, only the portion of the context "in a harbor" will change, with the remainder staying the same.

3.3.4. *Context-based intelligent assistant systems*

An important kind of intelligent system is the intelligent assistant system (IAS). These are AI systems that operate in conjunction with a human to accomplish tasks, for example, systems that handle subway incidents (Pasquier et al. 2000), medical diagnostic assistants (e.g. Attieh et al. 2013) and systems that help design pedagogical activities (Macchia and Brézillon 2021). Here, we are concerned with the requirements for those IASs that use context as a central component of their behavior, *context-based intelligent assistant systems* (CIASs).

CIASs, like all context-based systems, need explicit context representation, a way to recognize the context, and the means to transition between contexts. However, here, the operational definition of context must include the user and the system's relationship to them, and the task portion of the context is the task being accomplished by the user, with help from the system.

In contrast to an autonomous agent, in a CIAS, the user is ultimately the decision-maker. The system supports user decision-making by proposing satisfying alternatives and arguing for them, including recalling relevant points from previous interactions. The system needs to have a model of the process being considered and of the user and how they work (Brézillon 2014). Such objectives suppose a robust conceptual framework for modeling and managing context, and a computational tool for representing pieces of knowledge, reasoning and contexts in a uniform way (see Chapter 2). The three approaches discussed in Chapter 1, CxG, CxBR and CMB, offer excellent possibilities for the design and development of CIASs.

3.3.4.1. *Purpose of CIASs*

As an assistant, a CIAS is a decision-support system for users that have a high level of expertise in a domain which is not well known nor too complex. Users' expertise is highly compiled chunks of contextual knowledge mainly built by experience. Such expertise is generally used in a decision-making process leading to a critical and definitive decision.

There are several consequences of this for CIASs. First, the CIAS must be able to follow along with what the expert is doing and how they are doing it and anticipate their needs. This means that the system must have some model of the experts' reasoning. The system also needs to be a competent secretary, fixing all of the simple problems the experts encounter by itself and preparing dossiers on complex situations to enable the experts to make their own decisions.

Second, the CIAS needs to work from practices developed by the experts, with all of the contextual elements used by the expert during practice development. Thus, the system's line of reasoning must be based on the experts, as represented in the practices in its experience base. This gives the system a user-centered representation of the domain (Brézillon 2003).

Third, the system must develop a decision-making process in real time in order to analyze the diagnosis of problems and the actions taken by experts (Brézillon and Pomerol 1999). Indeed, the system must simultaneously develop the decision-making process and its own context-specific model (i.e. the practice).

Fourth, since the decision-making process is highly contextual, the system must take advantage of its interaction with the expert in order to learn new practices by incrementally acquiring any missing knowledge, thus enriching its experience base.

Fifth, by making context explicit in the experience base, the system has the possibility of generating relevant, context-grounded explanations. For example, it can generate relevant explanations for presenting the rationale behind a practice, along with abandoned alternatives; training (future) experts on the different practices developed; facilitating experience sharing among experts in a kind of dynamic corporate memory; and allowing a first step toward the certification of the user's protocol.

Sixth, the main tool of a CIAS must be a simulator that depends on a context-based representation. Indeed, such a simulator is the key element of a CIAS that does real-time decision-making because it is then possible to account for unpredicted events, thanks to explicitly modeling context as contextual elements that cover the user, the task realization, the working situation and the local environment with its available resources. All of the items are interdependent and also time-dependent. This assures that all contextual elements can be instantiated at the time when they are needed, and all contextual elements are taken into account at the moment when necessary.

3.3.4.2. *Abilities needed*

In addition to the abilities needed by any context-based intelligent system, a CIAS also requires several other abilities.

Simulation. As mentioned, a CIAS needs simulation abilities to allow the anticipation of outcomes, even when there is a lack of comprehensive a priori planning. The system determines how to behave according to the predictions, that is, anticipated future states that may affect its present states. A CIAS uses simulation to explore different dynamics of reasoning: validation, checking alternatives, replaying a reasoning sequence in different contexts, explanation, etc. Simulation provides explanations for the predictions, allowing the CIAS to be an effective decision-support system.

Explanation. Explanation is a means of developing a shared context between the CIAS and the user, with both contributing to its construction. In order to align the user's and the system's reasonings, both co-construct the explanation in the current context of problem-solving. This leads to cooperative problem-solving.

A CIAS interacts with agents that are experts in their domain; thus, the knowledge in the experience base is expert knowledge. Explanations concern the context of using the knowledge. For example, a proceduralized context in CxG is an ordered sequence of instantiated contextual elements. It can be used for explanation generation. Different categories were identified in CxG (see Brézillon 2003): visual explanations, dynamic explanations, user-based explanations, context-based explanations, micro- and macro-explanations, real-time explanations, etc. Moreover, these different types of explanation can be combined in different ways, such as visual and dynamic explanations for presenting future alternatives and abandoned options.

Knowledge acquisition. A CIAS cannot have all of the knowledge needed at design time, and so it needs to incrementally acquire new knowledge in its context of use. This approach relies on the fact that agents' knowledge is essentially a justification (i.e. explanation) of their judgment in that context.

Learning. The user of a CIAS is generally an expert with a high level of competence, for example, a specialist in medicine, such as an oncologist who makes critical decisions about a patient's cancer diagnosis and treatment. A learning tool is particularly important for such a situation, for example, when there is a discrepancy between a practice as it develops and the system's expectations based on its experience base. This may occur because the expert selects a different instantiation for a contextual element that has kept the same value during all practice development prior to some point.

A CIAS must learn by interacting with its users and becoming increasingly familiar with them and their ways of accomplishing their activities. Thus, CIASs must incrementally develop an experience base by assimilation (e.g. addition of variants) and accommodation (e.g. addition of a contextual element and new actions conditioned on them).

As a simple example, consider Figure 3.3. Initially, in part (a), action 1 is the usual way the CIAS knows what to do for a portion of a task, perhaps adding tap water to a coffeemaker. If the human expert decides to do a new action 2 instead (e.g. add distilled water instead of tap water), then the CIAS must acquire the new piece of knowledge (action 2) and create a new contextual element, as shown in Figure 3.3(b), with different values that explain why each branch is chosen and record this new practice in the experience base. This will cause the CIAS to recommend whichever action is appropriate in the future based on the context.

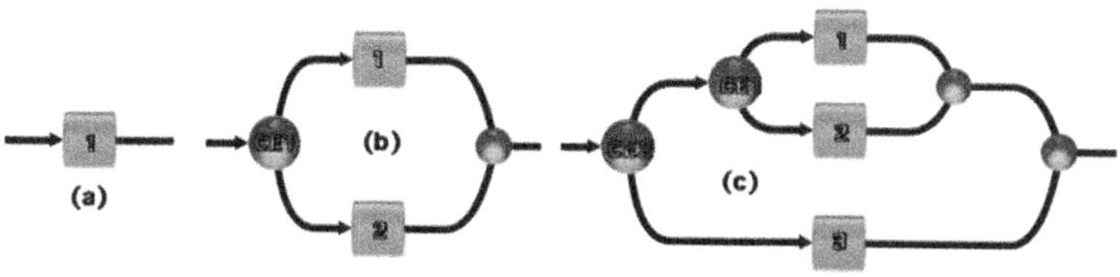

Figure 3.3. *Evolution of a contextual graph. For a color version of this figure, see www.iste.co.uk/brezillon/context.zip*

Now, if another agent has a completely different way to accomplish the task that is neither action 1 nor 2, nor is differentiated from them by a value of CE1 (e.g. a new actor makes coffee by heating tap water first to speed up coffee making), then the system must first learn CE2, the implicit context, put the previous graph on a branch of CE2 and add the new action 3 on another branch, as shown in Figure 3.3(c). Later, the system may help another user in the same task "coffee preparation" or introduce a new variant (e.g. use mineral water instead of tap water).

Communication. Communication is mandatory for user-CIAS interaction during their joint problem-solving. The communication component must be able to handle interruptions, reorient as needed, resume work at any point and possibly obtain information on the status of the task (e.g. Where are we? What can be done from this state?).

3.3.4.3. *Implementation framework*

To be in actual use for real-world tasks, a CIAS needs an editor with the usual functions for managing an experience base, reading practices, browsing alternatives of the practice that may be developed, exploring the experience base at different levels of granularity, analyzing (contextual) information attached to each item (date of creation, author, comments, etc.) and managing data.

In addition, to be useful and adopted, a CIAS should be written in a very common computer language (e.g. Python at the time of writing), must support different platforms (e.g. laptop, mobile, Web), and have its experience base stored as files in formats (XML, JSON, etc.) that allow them to be processed by other software. The design and development of the system must be user-centered for easy use by non-specialists in computer science and mathematics, and a short training time must be sufficient to allow non-computer scientists to start modeling a problem. The system should also be able to link to/export documents in common formats (e.g. word, PDF) and run other software as needed.

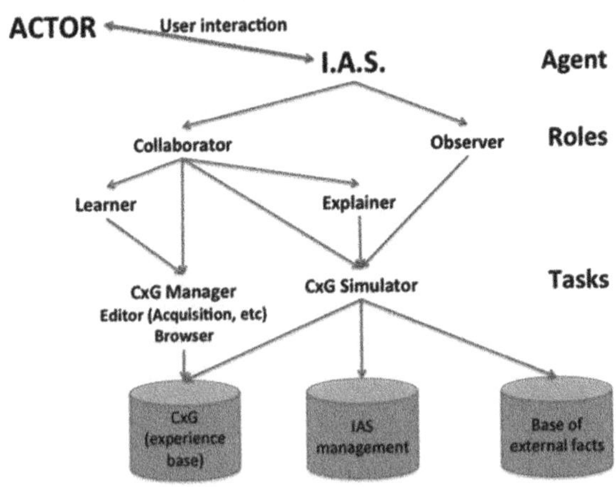

Figure 3.4. *General architecture of a CIAS*

CIAS must also support browsing. Browsing differs from simulation with respect to time in the decision-making process. A browser works at a tactical level where time is not considered explicitly (e.g. the duration of an action does not matter); it is more concerned with a qualitative comparison rather than the effective development of the task in a specific context. The focus is more on the realizability of the task than its effective realization (i.e. the first step in the decision-making process). For example, in a CIAS used in the Paris subway (Pomerol et al. 2002), operators used CxG-based browsing to replay how a colleague solved a given incident, in order to look for alternatives, analyze the working contexts and even study new practices.

Figure 3.4 shows the general architecture of a CIAS using the contextual graph (CxG) framework. In addition to the tools already discussed, there is a base of external facts that allows the transfer of some data or information between the different external applications and services triggered by an action in the contextual graph. The CIAS manages this fact base because some information at the operational level may be relevant for it (e.g. a change in the instantiation of a contextual element). Domain expertise is contained uniquely in the experience base, and data and information about the domain are contained in the base of external facts. Thus, this architecture may be applied in different domains by changing the knowledge bases.

3.3.5. *Context-based intelligent autonomous agents*

Another major class of intelligent systems are autonomous agents, both hardware-based agents (i.e. robots) and software-based agents (e.g. softbots, chatbots, etc.). Intelligent autonomous agents can either be completely autonomous or participants in multiagent systems.

While some of the requirements for a context-based intelligent autonomous agent (CIAA) overlap those of CIASs, some are different or are needed for different reasons.

3.3.5.1. *Purpose of CIAAs*

CIAAs have application wherever intelligent agents are needed, for example: controllers for autonomous automobiles (e.g. UCF's entry in the 2007 DARPA Grant Urban Challenge; Patz et al. 2008) and other land robots, uncrewed air vehicles, autonomous underwater vehicles, satellites and other space vehicles; process control and factory automation; software applications such as intelligent search engines, softbots and intelligent personal assistants; intelligent experiment controllers; and a host of military and healthcare applications.

A context-based intelligent autonomous agent differs from a CIAS in that it is not inherently a collaborative system, and so it does not need to focus as much on interacting with humans. It, unlike a CIAS, is ultimately the decision-maker. It will generally have its own goals that it is responsible for carrying out. Some of these will of course come from its human users and be directly mission-related, for example, "map area". Sub-goals instrumental to these will arise based on how the agent chooses to achieve the user goals. Still other goals will be mission-independent, for example, "avoid harming humans" or "survive". Many of these will be preservation goals (Schank and Abelson 1977), while others will be more active (e.g. "send status report to base").

Given this, a distinguishing characteristic of a CIAA is having a means to reason about and achieve goals, what is often called a "performance element" (e.g. Russell and Norvig 2020). Since the primary purpose of the agent is to solve problems autonomously, the performance element is not constrained to be human-like or even human-understandable, although the latter is helpful for knowledge acquisition and explaining its actions to humans. There is a wide range of techniques for performance elements, and they largely define the class of the CIAA. Examples include heuristic search agents, planning agents, rule-based agents and deep learning (usually deep-reinforcement learning) agents.

CIAAs operate over widely differing time scales depending on what their purpose is. For example: a UAV (unoccupied aerial vehicle or autonomous drone) tasked with flying over an area and returning video; a ground vehicle competing in DARPA's Grand Urban Challenge (Patz et al. 2008); an autonomous underwater vehicle operating for a week or more to sample data over a large (>1,000 km^2) area (*HUGIN Endurance*, n.d.); a solar-powered AUV staying deployed for weeks or months and returning data (Jalbert et al. 1997); a Mars rover operating for months or years; or a software agent monitoring a network for attacks over an unbounded amount of time. Thus, in addition to operational knowledge (what to do *right now*), they also need tactical knowledge (how to achieve overarching goals) and strategic knowledge (the larger, usually long-term goals).

This includes not only the knowledge directly used for achieving the agent's goals, but also contextual knowledge at different scales. For example, consider an intelligent controller for an autonomous underwater vehicle tasked with monitoring a coastal area for pollution. At any given time, it may need to know about the strategic context (the overall mission, which may span months), the current tactical context (e.g. finding the source of pollution that has been detected) and the operational context (e.g. maneuvering through a kelp bed while following the pollution odor plume to its source). Some approaches have strictly carved up these levels (e.g. CxBR, CMB as implemented in the MEDIC reasoner), while others (e.g. CMB in general) have not, rather allowing the granularity of the context and its associated knowledge to be determined by the needs of that particular context.

3.3.5.2. *Abilities needed by a CIAA*

Sophisticated context assessment ability. CIAAs usually operate in open, dynamic worlds in which the operational, tactical and strategic contexts can change rapidly and without warning. Failing to properly identify the current context for any CIAA risks mission failure; for physical autonomous agents (e.g. robots), the consequences for the agent itself can be dire, up to and including the loss of the agent. And for some agents (e.g. an autonomous automobile, a missile or rocket controller), the cost can include real harm to people or infrastructure.

Unlike the case with context-aware applications, where what is considered to be the context is rather simple, a context-based real-world agent often faces a very complex, rapidly changing situation, a tremendous number of features that could indicate what the context is, and a high level of uncertainty (e.g. from sensor limitations, effector imprecision, lack of knowledge, presence of other agents, etc.). Consequently, it is important both that the agent be endowed with sophisticated context assessment capabilities and that it devotes whatever cognitive resources are necessary to monitor and diagnose the evolving context. CMB provides one example of sophisticated context assessment based on medical diagnosis, as described above.

Context assessment is even harder for CIAAs working in multiagent systems (MASs), since in addition to the normal physical and other features of the situation, the context now includes other agents whose motivations, reasoning styles, knowledge and capabilities are opaque to the agent. It is worth the effort, however. Knowing the context can allow an agent to understand what organizational roles it is expected to have based on what it knows of the context. For example, if it identifies a contracting context (perhaps by overhearing messages about bids), then it can use knowledge about that context to understand that it should bid for tasks, announce subcontracts and so forth.

Agents that are part of a cooperative MAS (e.g. a Cooperative Distributed Problem-Solving (CDPS) system; Lesser and Corkill 1988) can benefit from cooperatively developing an idea of what the joint context is and determining how their own local context fits into it. One use for this that has been explored (Rode and Turner 2015) is to use knowledge of the joint context to cooperatively decide what is the best organization (contract network, hierarchy, team, etc.) for the agents in the context.

Although at first glance it may seem that constraints on the response time of an autonomous agent would limit context assessment, this is not the case in many applications. While it is true that some agents have to operate in the real-time or soft-real-time regime (e.g. controllers for cruise missiles, launch vehicles or spacecraft during re-entry and landing), for others, this is not an issue due to the disparity between the speed of the computer system and that of the real world. For the AUV domain, for example, decisions about what to do typically do not require real-time performance simply because things happen slowly underwater for the most part. This is true even in domains where things happen faster, for example, controlling a UAV, since in general, things that *do* require real-time responses are often quite simple (obstacle avoidance in some cases) and can be handled by "lower-level" computer programs, while the intelligent agent reasons about longer-term, more complex tasks. To ensure context-appropriate fast responses, it is also possible, indeed advisable, for the CIAA to "preset" lower-level behaviors based on its contextual knowledge (see Chapter 5).

Simulation. While not all autonomous agents have or require it, simulation ability, at least in the form of plan projection, is beneficial for CIASs in order to allow the reasoner to choose the best course of action in a situation. Some of this can be aided by direct knowledge of the context. For example, in the context of entering a minefield, a robot should be able to easily predict outcomes arising from not checking for mines before moving.

Wilson (see Chapter 7) has extended this idea to *context* projection: determining, based on goals, possible plans and the current context, what future contexts the agent can expect to be in. It can then use knowledge about those possible future contexts to choose actions in the current context.

A CIAA that is part of a MAS will likely need the ability to model other agents. Multiagent system approaches usually require some type of modeling of other agents present, even if the model is quite simple, such as in game-theoretic approaches, where agents are modeled as their utilities, or in contracting approaches (e.g. Smith 1980), where they are modeled by their adherence to the contracting protocols. More complex agents, or *open systems* where the MAS composition can change over time in unexpected ways, require more sophisticated models. Here, a CIAA can leverage its contextual knowledge. For example, even if the agent has no deep model of another agent's beliefs, etc., there may be features of the other that allows it to recognize a context in which it has interacted with the other or similar agents in the past, and so bring context-specific knowledge about how the other agent will behave or react into focus. There have been context-based approaches to trust, for example, that do this (Whitsel and Turner 2015).

Explanation. While it may not be at first apparent, for many CIAAs, an explanation ability will be very important. One reason for this is user acceptance. A CIAA, especially the one whose actions could cause harm, is unlikely to be trusted by its users if it cannot explain to them why it did what it did. Explaining what it is *about* to do is equally important. Turner (see Chapter 5) observed, when working with military personnel, that they were not sanguine about the prospect of completely autonomous AI-controlled AUVs, precisely because the AI's behavior could not be accurately predicted and explained ahead of time.

Explanation is especially important when something goes wrong so that the humans responsible for the agent can understand the issue and take steps to solve the problem for the agent's future behavior. It is also important in some MASs. For example, in the approach known as partial global planning (PGP; Durfee and Lesser 1987), abstracted versions of agents' plans are exchanged to agree on a plan spanning larger groups of agents; the partial global plans are a kind of inter-agent explanation. Explanation for CIAAs is aided by contextual knowledge, since it clusters actions and rationale for them in a manner natural for humans to grasp.

Knowledge acquisition. For the foreseeable future, most of a CIAA's knowledge will, initially at least, come from humans. Consequently, they will need some way to acquire knowledge. This has implications for the kinds of knowledge representation they use, since it implies that the knowledge structures support and make obtaining knowledge efficient. Context-based systems have an advantage in this regard, since humans naturally understand the concept of a context and often have a great deal of information readily available about different contexts.

Gonzalez et al. (2005) describe a method for knowledge acquisition for CxBR, called context-based interactive tactical knowledge acquisition (CITKA). CITKA exploits the context-based nature of CxBR to semi-automatically acquire knowledge from a subject matter expert (SME). It asks the SME about contexts relevant to the mission and domain, then uses those to guide the acquisition of contextual knowledge and transition rules for context switching. CITKA was implemented in a proof-of-concept system, and the authors estimated that it saved 60% of the knowledge engineer's time and effort during knowledge acquisition.

Learning. A learning component may or may not be important for a context-based autonomous agent. For some applications, the missions may be short enough that it is sufficient for humans to alter the agent's knowledge between missions. Other missions, though, may be long enough that within-mission changes are needed; if communication is an issue, either due to low bandwidth (e.g. underwater, on another planet) or because the agent must remain silent and undetected (surveillance missions, various military missions), then the agent can benefit from incorporating learning. Such agents may need to learn autonomously, which necessitates machine-learning capabilities such as (symbolic or neural network-based) induction, explanation-based learning, case-based reasoning or reinforcement learning (deep or otherwise). Any learning technique used will need to be capable of learning in context and needs to modify the agent's contextual knowledge, set of represented contexts, context-assessment knowledge and/or the organization of its contextual knowledge.

Communication. Depending on its purpose and tasks, a CIAA is likely to need at least *some* communication ability. For long-duration CIAAs, status reports and interim results, etc., may be expected by its users. Agents that find themselves in situations they cannot handle may need to communicate with humans to obtain help.

CIAAs can bring their contextual knowledge into focus in several ways to aid communication. This can include deciding the timing of communication. If the agent knows it is in a context where communication is impossible (no communication channel currently available, for example) or unwise (in hostile territory), then it can quickly know to not attempt it, but rather to postpone communicating until in a more favorable context. Often, real-world agents have multiple possible communication channels available to them, each with different properties; for example, an AUV may have acoustic communication, radio and/or even laser communication abilities. Contextual knowledge can help the agent quickly and accurately decide which to use (e.g. acoustic communication for long-range underwater communication, radio when surfaced, laser when near the other conversant in clear water, etc.).

Agents in MASs need to know when to communicate, what to communicate and how. Context can also help here, especially knowledge about the cooperation protocols in effect in the current situation. Communication is also needed to allow agents to develop a notion of the joint context. These agents need a communication language that can not only express contextual knowledge, but also refer to specific, mutually known contexts, as well as contexts that are idiosyncratic to themselves. While having a particular language is not overly important and may change, depending on the context (low- vs. high-bandwidth contexts, for example), an *ontology* for contextual knowledge and contexts themselves is needed. This allows the agents to cooperatively determine their joint context, as well as potentially identify faulty knowledge, including faulty context assessment, by their peers. An initial ontology for CMB, for example, can be found in the work by Rode and Turner (2015).

3.3.5.3. *Implementation framework*

The diversity of uses for CIAAs means that there is not a single implementation framework that will work for all of them. At minimum, all will require some way of giving them their initial knowledge, including contextual knowledge, and a description of their mission or goals. Beyond that, though, their purpose will drive how they are implemented.

CIAAs that control hardware will need some way to interact with the hardware. This could be directly (e.g. by sending signals to particular hardware ports), but will likely be via interacting with more specialized, faster control software. Some applications can benefit from using a standard middleware, for example, the Robot Operating System (ROS; Quigley et al. 2009) or one of many homegrown varieties (e.g. the EAVE architecture, Blidberg and Chappell (1986), or the Knight Rider VIP's architecture, Patz et al. (2008)). For those that are software agents (e.g. Turner et al. 2021), they will usually interact with their host computer using procedure calls while interacting with other software, locally or otherwise, via agreed-upon high-level protocols and underlying Internet protocols.

Which information CIAAs send and receive, as well as commands they send, also depends on the application. Some will receive sensor data, either "raw" data or data that has been processed into what we referred to in Figure 1.2 as context-dependent information and/or knowledge. Information sent to others – users or other CIAAs – will generally be at the knowledge level or constitute commands sent to lower-level software components.

CIAAs may operate with or without an "operator in the loop", depending on their missions. For those that must interact with humans during their missions, some sort of (minimal) NLP or GUI will be needed. For those CIAAs operating with other agents, for example, in a MAS, then agreed-upon cooperation protocols and languages will be needed.

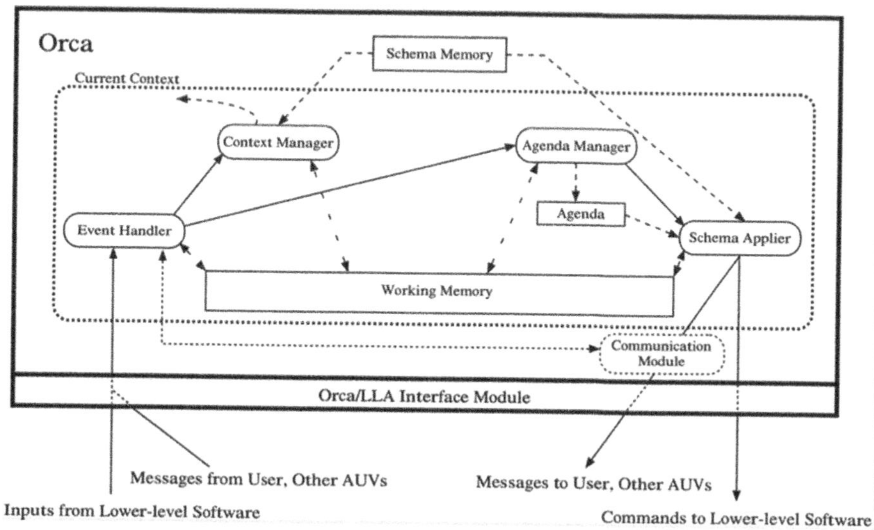

Figure 3.5. *An example CIAA architecture from Turner's work*

With respect to the agent's own software architecture, there is no single agreed-upon architecture. Some might be based on some standard middleware, for example, the Robot Operating System (ROS; Quigley et al. 2009). Others will be homegrown. An illustrative example is shown in Figure 3.5. This is the internal architecture of one CIAA, Orca (see Chapter 5).

3.4. Conclusion

The tension in the field of AI has always been between, on the one hand, formal theories (e.g. theorem proving rooted in formal logic) and system engineering and, on the other hand, appeals to "softer" theories arising from cognitive science (e.g. means–ends analysis) and pragmatic, exploratory approaches.

However, technology evolution is seldom associated first with formal models. Usually, there is a pragmatic need to solve a problem that produces a particular, pragmatic solution (often, of course, informed by formal methods). This is as true for context-based reasoning as any other kind.

Over time, there has also been a growing realization that using contextual knowledge is necessary to solve complex problems, and that it is most beneficial to explicitly represent (model) for use by reasoners. In formal approaches (e.g. theorem proving), contexts have usually been assumed to be discrete entities with little to no intrinsic structure that could be exploited by a reasoner. In more pragmatic approaches (e.g. rule-based systems), the need for context was clear, but the use of contextual knowledge was initially implicit, with contextual knowledge relegated to rule antecedents, for example.

However, making context explicit has many benefits. It helps an agent to know what context it is in, so it can use its knowledge and behave appropriately for the context. It allows capturing context-specific variants of reasoning (e.g. practices developed by operators or by itself in the past). It allows explanations to be made in context, which facilitates understanding. It facilitates knowledge acquisition, since humans are naturally context-aware. And it allows inferences to be made about contexts, their relation to each other and how to transition from one to another.

Formally, context is difficult to capture, as McCarthy (1993) pointed out: it is strongly interrelated to domain knowledge and reasoning, it is recursive, and it has an infinite dimension. Determining which facts to include in a context is a variant of the frame problem (Shoham 1987).

Nevertheless, pragmatic approaches persevered, pointing out that from a pragmatic perspective, context is what constrains a reasoner's problem-solving without directly intervening. Thus, only the part of the context that is useful needs to be modeled. As a result, pragmatic approaches in AI emphasize their systems' ability to recognize the context they are in, make decisions and behave appropriately in real-world domains, even without taking into account the context's problematic infinite dimension, recursion and susceptibility to the frame problem.

This chapter attempts to pave the way for more context-based intelligent systems by discussing the requirements for their design and implementation. In addition to the three major things required of all such systems – explicit context representation, context-assessment ability and a means to transition between contexts – we also discussed the particular requirements for two important classes of context-based intelligent systems, context-based intelligent assistant systems (CIASs) and context-based intelligent autonomous agents (CIAAs).

Doubtless, this set of requirements is incomplete. Nevertheless, we hope that this work will frame future discussions and perhaps provide a kind of roadmap for future research on one of the most important areas of AI: pragmatic, context-based intelligent systems.

3.5. References

Aikins, J.S. (1980). Prototypes and production rules: A knowledge representation for computer consultations. Master's Thesis, Stanford University, Department of Computer Science, Stanford.

Attieh, E., Brézillon, P., Capron, F. (2013). Context-based modeling of an anatomo-cyto-pathology department workflow for quality control. In *Modeling and Using Context (Proceedings of the 2013 International and Interdisciplinary Conference on Modeling and Using Context)*, Brézillon, P., Blackburn, P., Dapoigny, R. (eds). Springer, Berlin, Heidelberg.

Blidberg, D.R. and Chappell, S.G. (1986). Guidance and control architecture for the EAVE vehicle. *IEEE Journal of Oceanic Engineering*, 11(4), 449–461.

Bobrow, D.G. and Brady, J.M. (1998). Artificial intelligence 40 years later. *Artificial Intelligence*, 103(1–2).

Brézillon, P. (1989). Interpretation and rule packet in expert systems. Application to the SEPT expert system. *International Conference on Knowledge-Based Computer Systems*, 78–87.

Brézillon, P. (2003). Focusing on context in human-centered computing. *IEEE Intelligent Systems*, 18(3), 62–66.

Brézillon, P. (2014). A context-centered architecture for intelligent assistant systems. In *Innovations in Intelligent Machines*, Faucher, C. and Jain L.C. (eds). Springer, Berlin, Heidelberg.

Brézillon, P. and Abu-Hakima, S. (1995). Using knowledge in its context: Report on the IJCAI-93 workshop. *AI Magazine*, 16(1), 87–87.

Brézillon, P. and Gonzalez, A.J. (2006). Tale of two context-based formalisms for representing human knowledge. In *Advances in Applied Artificial*, Ali, M. and Dapoigny, R. (eds). Springer, Berlin, Heidelberg.

Brézillon, P. and Pomerol, J.-C. (1996). Misuses and nonuse of knowledge-based systems: The past experiences revisited. In *Implementing Systems for Supporting Management Decisions*, Humphreys, P., Bannon, L., McCosh, A., Migliarese, P., Pomerol, J.-C. (eds). Chapman and Hall, London.

Brézillon, P. and Pomerol, J.-C. (1999). Contextual knowledge sharing and cooperation in intelligent assistant systems. *Le Travail Humain*, 62(3), 223–246.

Brézillon, P., Pasquier, L., Pomerol, J.-C. (2000). Representing operational knowledge by contextual graphs. *Advances in Artificial Intelligence, International Joint Conference*, 245–258.

Chandrasekaran, B., Gomez, F., Mittal, S., Smith, J. (1979). An approach to medical diagnosis based on conceptual structures. *Proceedings of the Sixth International Joint Conference on Artificial Intelligence*, 134–142.

Clancey, W.J. (2002). Simulating activities: Relating motives, deliberation, and attentive coordination. *Cognitive Systems Research*, 3(3), 471–499.

Cullingford, R.E. (1981). SAM. In *Inside Computer Understanding*, Schank, R.C. and Riesbeck, C.K. (eds). Lawrence Erlbaum Associates, Mahwah.

Davis, M. and Putnam, H. (1960). A computing procedure for quantification theory. *Journal of the ACM (JACM)*, 7(3), 201–215.

Durfee, E.H. and Lesser, V.R. (1987). Using partial global plans to coordinate distributed problem solvers. *Proceedings of the 1987 International Joint Conference on Artificial Intelligence*, 875–883.

Fikes, R.E. and Nilsson, N.J. (1971). STRIPS: A new approach to the application of theorem proving to problem solving. *Artificial Intelligence*, 2(3–4), 189–208.

Giunchiglia, F. (1993). Contextual reasoning. *Epistemologia, Special Issue on I Linguaggie Le Macchine*, 16, 345–364.

Gonzalez, A.J., Castro, J., Gerber, W. (2005). Automating the acquisition of tactical knowledge for military missions. *The Journal of Defense Modeling and Simulation*, 2(3), 145–160.

Gonzalez, A.G., Stensrud, B.S., Barrett, G. (2006). Formalizing context-based reasoning: A modeling paradigm for representing tactical human behavior. *International Journal of Intelligent Systems*, 23, 822–847.

HUGIN Endurance (n.d.). Konsberg [Online]. Available at: https://www.kongsberg.com/maritime/products/marine-robotics/autonomous-underwater-vehicles/autonomous-underwater-vehicle-hugin-endurance [Accessed 27 November 2021].

Jalbert, J.C., Irazoqui-Pastor, P., Miles, S., Blidberg, D.R., James, D. (1997). Solar AUV technology evaluation and development project. *Proceedings of the Tenth International Symposium on Unmanned Untethered Submersible Technology*, 75–87.

Kolodner, J.L. (1987). Capitalizing on failure through case-based inference. *Proceedings of the Ninth Annual Conference of the Cognitive Science Society*, 715–726.

Laird, J.E., Newell, A., Rosenbloom, P.S. (1987). SOAR: An architecture for general intelligence. *Artificial Intelligence*, 33(1), 1–64.

Lesser, V.R. and Corkill, D.D. (1988). Functionally accurate, cooperative distributed systems. *Readings in Distributed Artificial Intelligence*, 295–310.

Lindsay, R., Buchanan, B.G., Feigenbaum, E.A., Lederberg, J. (1980). *Applications of Artificial Intelligence for Chemical Inference: The DENDRAL Project*. McGraw-Hill, New York.

Macchia, G. and Brézillon, P. (2021). Automated generation of pedagogical activities adapted to the learning context. *Modeling and Using Context*, 21(4) (CONTEXT-21 Special Issue) [Online]. Available at: http://openscience.fr/IMG/pdf/iste_muc21v4n1_7.pdf.

McCarthy, J. (1993). Notes on formalizing context. *IJCAI*, 555–560.

McCarthy, J. and Hayes, P.J. (1969). Some philosophical problems from the standpoint of artificial intelligence. In *Machine Intelligence 4*, Meltzer, B. and Michie, D. (eds). Edinburgh University Press, Edinburgh.

Miller, R.A., Pople, H.E., Myers, J.D. (1982). INTERNIST–1, an experimental computer-based diagnostic consultant for general internal medicine. *New England Journal of Medicine*, 307, 468–476.

Minsky, M.L. (1975). A framework for representing knowledge. In *The Psychology of Computer Vision*, Winston, P.H. (eds). McGraw-Hill, New York.

Newell, A. and Simon, H.A. (1963). GPS: A program that simulates human thought. In *Computers and Thought*, Feigenbaum, E.A. and Feldman, J. (eds). McGraw-Hill, New York.

Newell, A. and Simon, H.A. (1976). Computer science as empirical inquiry: Symbols and search. *Communications of the ACM*, 19(3), 113–126.

Pasquier, L., Brézillon, P., Pomerol, J.-C. (2000). From representation of operational knowledge to practical decision making in operations. *Decision Support through Knowledge Management*, 301–320.

Patz, B.J., Papelis, Y., Pillat, R., Stein, G., Harper, D. (2008). A practical approach to robotic design for the DARPA urban challenge. *Journal of Field Robotics*, 25(8), 528–566.

Pohl, F. (1969). *The Age of Pussyfoot*. Ballantine, New York.

Pomerol, J.-C., Brézillon, P., Pasquier, L. (2002). Operational knowledge representation for practical decision-making. *Journal of Management Information Systems*, 18(4), 101–115.

Quigley, M., Conley, K., Gerkey, B., Faust, J., Foote, T., Leibs, J., Wheeler, R., Ng, A.Y. (2009). ROS: Open-source robot operating system. *ICRA Workshop on Open Source Software*, 3, 5.

Rode, S. and Turner, R.M. (2015). Representing and communicating context in multiagent systems. In *Modeling and Using Context (Proceedings of CONTEXT-05)*, Christiansen, H., Stojanovic, I., Papadopoulos, G.A. (eds). Springer, Cham.

Russell, S. and Norvig, P. (2020). *Artificial Intelligence: A Modern Approach*, 4th edition. Pearson, London.

Schank, R.C. (1983). *Dynamic Memory: A Theory of Reminding and Learning in Computers and People*. Cambridge University Press, Cambridge.

Schank, R.C. and Abelson, R. (1977). *Scripts, Plans, Goals and Understanding*. Lawrence Erlbaum Associates, Mahwah.

Shoham, Y. (1987). What is the frame problem? In *The Frame Problem in Artificial Intelligence*, Brown, F.M. (ed.). Morgan Kaufmann, San Mateo.

Shortliffe, E.H. (1974). MYCIN: A rule-based computer program for advising physicians regarding antimicrobial therapy selection. PhD Dissertation, Department of Computer Science, Stanford University.

Smith, R.G. (1980). The contract net protocol: High-level communication and control in a distributed problem solver. *IEEE Transactions on Computers*, C–29(12), 1104–1113.

Smith, R.G. (1985). *Knowledge-Based Systems: Concepts, Techniques, and Examples*. Schlumberger-Doll Research, Cambridge.

Sticklen, J. (1987). MDX2: An integrated medical diagnostic system. PhD Dissertation, Department of Computer and Information Science, The Ohio State University.

Toosi, A., Bottino, A.G., Saboury, B., Siegel, E., Rahmim, A. (2021). A brief history of AI: How to prevent another winter (a critical review). *PET Clinics*, 16(4), 449–469.

Turner, R.M. (1994). *Adaptive Reasoning for Real-World Problems: A Schema-Based Approach*. Lawrence Erlbaum Associates, Mahwah.

Turner, R.M. (1998). Context-mediated behavior for intelligent agents. *International Journal of Human–Computer Studies*, 48(3), 307–330.

Turner, R.M. (2014). Context-mediated behavior. In *Context in Computing: A Cross-Disciplinary Approach for Modeling the Real World Through Contextual Reasoning*, Brézillon, P. and Gonzalez, A.J. (eds). Springer, New York.

Turner, R.M., Loftin, C., Revello, A., Kline, L.R., Lewis, M.A., Yasai Sekeh, S. (2021). Context-dependent deep learning. *Modeling and Using Context*, 21(4) (CONTEXT-21 Special Issue) [Online]. Available at: https://www.openscience.fr/IMG/pdf/iste_muc21v4n1_8.pdf.

Whitsel, L. and Turner, R.M. (2015). Using contextual knowledge for trust strategy selection. In *Modeling and Using Context (Proceedings of the Ninth International and Interdisciplinary Conference on Modeling and Using Context)*, Christiansen, H., Stojanovic, I., Papadopoulos, G.A. (eds). Springer, Cham.

Woods, D.D. (1985). Cognitive technologies: The design of joint human–machine cognitive systems. *AI Magazine*, 6(4), 86–92.

4

Activating the Context for Learning and Teaching: Findings from the TEEC Project

The last decade has seen a turning point in research on context effect-based education with the implementation of the TEEC project (*Technologies Éducatives pour l'Enseignement en Contexte* – Educational Technologies for Teaching in Context). This project is aimed at conceptualizing a techno-pedagogical model where context, in the various forms that it may take, occupies a central role. Several experiments using the context effect-based pedagogical approach have been completed successfully. The fundamental principle of this approach is the involvement of student groups living in distinct geographic areas, sharing the same language and collaborating on subjects that have local specificities. The collaboration is done synchronously and asynchronously through existing digital platforms. Multiple digital tools are being developed to assist in the creation of such pedagogical projects, including (1) Mazcalc: a context modeler and calculator that can be used to model an object of study in distinct geographical contexts. It is connected to (2) an authoring system that creates learning scenarios based on the contextual specificities of the study project in the two geographical areas involved. And (3) a context-aware intelligent tutoring system is designed to support learning, teaching and research activities. In this chapter, the global findings on the context effect-based education and research will be detailed along with the developments in digital tools.

4.1. Introduction

In this chapter, the primary objective is to summarize and compile the work progress, and results achieved within the TEEC project[1]. Within the general context of

---

Chapter written by Claire ANJOU, Thomas FORISSIER, Jacqueline BOURDEAU, Valéry PSYCHÉ, Lamprini CHARTOFYLAKA and Alain STOCKLESS.

context-based education, this project incorporates multiple research areas such as: science education including learning and teaching, educational digital technologies, computer-supported collaborative learning, image recognition, human–computer interaction, instructional design, intelligent tutoring systems and didactic contextualization. It aims to create an educational concept in which context, its effects and activation occupy a central position in the process of teaching and learning.

The TEEC project originated from a meeting between researchers from two laboratories: one in Quebec and one in Guadeloupe, in 2011. The founding members shared the belief that context is a crucial element in education and that it should not only be seen as a factor influencing teaching and learning, but also as an asset for better comprehension of these processes. A discussion between the researchers about the challenges of using and activating context for teaching led them to think about a pedagogical model that would emphasize the role of context in science education. The authors conceptualized that the creation of collaborative pedagogies exploiting the contrast between two geographic areas would provide a great opportunity to create a stimulating and meaningful education using the contexts, and it would also be a great opportunity to investigate the importance and diversity of contexts. Quebec and Guadeloupe were chosen as the testbed for their educational model since Guadeloupe, a French overseas territory, and Quebec, a Canadian province, have two very different climates, environments, cultures, politics, economics, legal systems and educational systems. Despite these differences, the two contexts share some commonalities, such as the time zone and language (despite variations in accents and influences) which could lead to a fruitful collaboration.

In order to have a better understanding of the text, semantic precision is crucial. A contrast between two contexts is called a "context gap" and is quantified to be calculated. The instructional model is known as the "CLASH model", to illustrate the benefit of this model by stimulating the awareness of students about the role of context. Instructional design is designed to promote this awareness and implemented in an instructional scenario that includes both sides.

The term "experiment" should not be understood within a framework of experimentalism but rather as a process of designing an educational situation and collecting data in the field. In design-based research (DBR) methodology, it is also called "iteration". Besides, the pioneers used the term "design experiments", (Brown 1992). It is therefore not experimentalism, which requires the control of variables, but testing the value and effectiveness of an instructional design when implemented in the field, with the intention of making it better and improving the data collection instruments through multiple iteration loops. Consequently, the term "effect" is not to be understood in the sense of a casual event, but as one that is triggered by the learning activities.

1 https://teec.teluq.ca/fr/.

Fundamentally, the TEEC project relies on the implementation of multiple experiments using the context effect-based pedagogical approach. As explained, this approach involves learners and teachers living in distinct geographical areas, teaching and learning collaboratively on common subjects with local or cultural specificities. As part of pedagogical experimentation, this project also intends to develop digital tools to assist and participate in the creation of such pedagogical projects. Three main products result from the development of these tools. The first one is (1) a context modeler and calculator: Mazcalc. It allows for modeling an object of study in distinct geographic contexts; it provides information about the differences and similarities between the various contexts' instantiations of a given object, based on several parameters and themes. Mazcalc is connected to (2) an authoring system: Scenarizator. This enables the creation of learning scenarios that are based on the contextual specificities of the studied object in both geographical areas. It aims to define the learning activity, the learning objectives, teaching and learning tasks, evaluation methods, as well as materials and numerical facilities. An important aspect of Scenarizator is its ability to organize the pedagogical and temporal synchronization of experiments. It is based on the pedagogical context and takes into account the pedagogical constraints, such as the institution, the curriculum and the time schedule. Finally, the last tool is (3) a context-aware intelligent tutoring system (CAITS) that supports the learning and teaching processes while providing assistance to the research activities by tracking data from the pedagogical experiments.

This chapter is organized as follows: first, theoretical references discuss the notion of context in education, context effects in education and digital technologies in education. In the second part of the chapter, the global research focuses on the TEEC project. The complex methodology of design-based research, data collection and analyses, and the global methodological organization of the project are discussed. The results are then organized according to each specific research dimension. Finally, the chapter concludes by describing the main advances in the conceptualization and development of the pedagogical model.

4.2. Theoretical framework

4.2.1. *Internal and external contexts for education*

In the field of education, the term "context" has many definitions, which often depend on the discipline in which it is studied. In education sociology, it refers to the institutional context and explains the relationships between contextual and individual parameters to achieve academic success (Duru-Bellat and Mingat 1988). In language education, the context concerns the didactic situation. It is generally accompanied by a command of language for learners, as well as the teacher's ability to adapt their teaching (Blanchet et al. 2009). In science education, context is generally described as the student's physical environment, and context-based approaches link the learning to real-world situations to allow students to better understand the possible applications of what they learn (Schwartz et al. 2004; Blanchet et al. 2009). In accordance with these approaches, field investigations and natural environment observations are examples of how context can be used to motivate and engage learners.

The term "context" is also very prominent in cognitive psychology work about knowledge (Bastien 1998; Bazire and Brézillon 2005). Cognitive processes are linked to situated knowledge and are context-dependent. The situated knowledge is organized in a functional way and activated by the context according to the goal to be accomplished. The relationship between context and cognition gives the context either an internal or external nature (Bazire and Brézillon 2005). In the first case, it belongs to the individual and occupies an important place in the representation that the individual constructs for themselves regarding the situation in which they are involved. However, in the other case, the context is related to elements external to the individual. The concept of authentic external and internal context is also well described by Van Wissen et al. (2013). They consider that the external context participates in the construction of learners' mental representations, that is, learners' internal context or learners' conceptions. We agree with Wertsch (1991) that the study of a learner's mental functioning is intrinsically embedded in the social, cultural, institutional, historical, as well as geographical and environmental external contexts. Internal and external contexts are either internal to the individual mind or external to it. In addition, we believe that both are highly dependent on each other and have complex relationships.

Although there are different definitions of the term context, both its internal and external natures are known to have a significant impact on learning and teaching (Anciaux 2013; Delcroix et al. 2015). Sometimes, both contexts can become an obstacle to knowledge acquisition if not taken into consideration. In fact, learners' internal contexts, that is, their conceptions, can be associated with misconceptions and influence cognitive learning processes. Similarly, external context, which is a driving force in the comprehension of scientific concepts and whose positive effects on authentic teaching are well known (Brown et al. 1989; Schwartz et al. 2004), can also impact learning if it does not match the conceptions held by the curriculum or textbooks, that is, the learning reference (Forissier 2015). Context effects are greatly affected by the distinction between external and internal contexts. Context effects were first described in Guadeloupe where the application of institutional curricula, developed in continental France, is often not adapted to local situations. They led to the development of multiple works on didactic contextualization and contextual teaching and learning (Delcroix et al. 2013). The term "context effect" refers to the phenomenon that occurs in a didactic situation in which a gap or a difference between multiple contexts is highlighted (between the internal contexts of two actors in a didactic situation, or between the internal context of a conception's owner and its external context). According to Forissier (2015), these phenomena, which are perceived as obstacles, can support learning and be used as tools for teaching. Toward this aim, he developed, together with colleagues from the TEEC, the CLASH model, which is a context effects-based pedagogical approach (Forissier et al. 2013). This method involves organizing pedagogical scenarios and experiments in which the encounter between learners with different internal and external contexts is at the center of the learning process. To do so, two groups of students living in different geographical locations are requested to perform simultaneous investigations in their local contexts on common subjects based on a common pedagogical scenario. Then, they are asked to collaborate so they can share their observations on common concepts through inter-team and inter-group exchanges. When collaborating, they are expected to share their conceptions and if a gap is highlighted then the CLASH effect or the context effect occurs. Through this effect, students come to realize how their conceptions are contextualized, paving the way for learning new ones. The first pedagogical experiments based on context effects were conducted with students living in Guadeloupe and Quebec in the field of science, focusing on objects possessing contextual specificities in each context (Fécil 2014; Anjou 2018). The study of these experiments revealed that context effects appear during interaction moments and lead to the emergence of emotions. According to this hypothesis, context effects can be used as an opportunity for learners to better understand diversity, enrich their conceptions and develop critical thinking skills.

4.2.2. *Modeling external context*

A study of context and digital technologies in education cannot be complete without mentioning artificial intelligence (AI). AI is crucial for developing adaptive digital tools with the purpose of facilitating learning and instruction through context modeling, knowledge modeling, recommending, coaching and tutoring. However, knowledge representation and modeling are complex tasks in the field of AI.

In the AI perspective, the need to model knowledge required for reasoning is the raison d'être of knowledge representation and modeling: reasoning about entities in the world requires a representation of these entities. According to the symbolic approach to AI, reasoning consists of using an algorithm to manipulate objects and relations expressed in symbolic form: inference is the result of this reasoning (Russell and Norvig 2010). Knowledge representations define the entities and relations included in a reasoning mechanism. The nature of a representation determines what is represented (what) and, from this viewpoint, this representation is strongly dependent on the cognitive domain or task to which the reasoning corresponds. Declarative representations have rich semantics due to the diversity of meanings supported by the associated form and due to the fact that they relate multiple objects. In general, the term semantic representation is associated with formal representations such as semantic graphs and ontologies. Semantic representations are distinguished by their qualitative characteristic: they define semantic relations between entities without necessarily attributing to them a truth value or a certainty value or a temporal dimension. They are conceptual graphs where the objects of the world represented correspond to the nodes, while the links (to which meaning is assigned) correspond to the relations between these objects. The links are oriented since the relationships are not necessarily symmetrical. With semantic representations, it is possible, among other things, to inherit object properties as well as handle exceptions, which are not possible with logical representation formalisms. Semantic networks (Sowa 1991), ontologies (Gruber 1993) and Bayes networks (Pearl 1988) are semantic representation formalisms. In all cases, concepts (object, action, event, process) are represented in the form of graphs composed of interrelated nodes. Formal knowledge representations are operationalized in knowledge bases. Knowledge bases are implemented in computers as data, which are then used by information management systems and AI algorithms. Ontologies are usually formalized in a semantic web language (XML, OWL, etc.) to be implemented in an AI system.

The TEEC project addressed this issue when it came to performing reasoning with context-aware tools (specifically with Mazcalc and its interaction with Scenarizator and CAITS). To achieve this goal, our approach turned to declarative representations, specifically ontologies. According to our community of researchers, ontological engineering is the most appropriate way to model the context in order to have a common, consensual vision of what it represents. Indeed, ontology engineering is used when not everyone agrees on the same characterizations of a domain of interest, which was the case for TEEC. Consequently, an ontology was used to reveal consensual knowledge about our vision of the external context for the TEEC project (Figure 4.1). The TEEC team used ontological modeling to understand theory about context and to articulate it into context-aware tool design models. Specifically, modeling an ontology of the context has allowed Mazcalc to explicitly specify the conceptualization of its databases. This ontology will be used in the same way as with other context-aware tools developed by TEEC. From a computer science perspective, it can be viewed as the backbone of the artifact's database. Ontological modeling forced the TEEC researchers to reflect deeply on a common understanding of the "context" they wanted to use to serve the project purpose, and its relationship to the didactic situation.

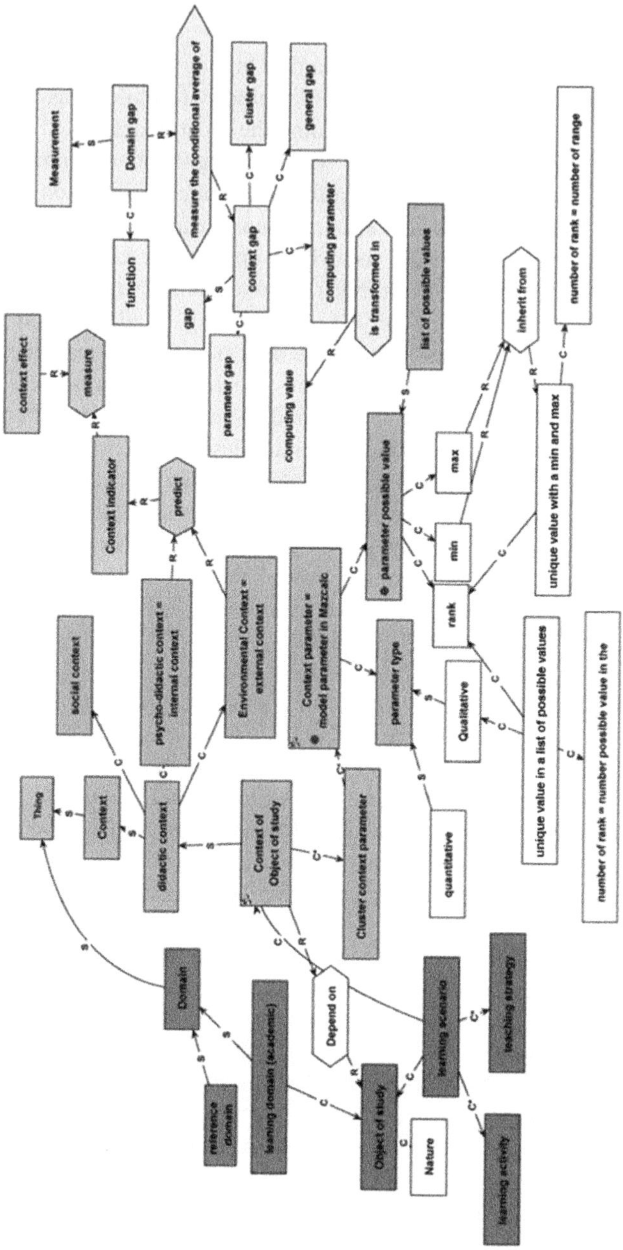

Figure 4.1. *External context ontology. For a color version of this figure, see www.iste.co.uk/brezillon/context.zip*

As explained in Psyché et al. (2003), the origins of Ontology (from the Greek, *onto logos*) can be traced back to Aristotle's metaphysics, whereas in the field of philosophy, the term ontology appeared simultaneously in its Latin form in 1613 from two philosophers, Göckel and Lorhard. Ontology was defined first by McCarthy (1980). Although Gruber's (1993) definition of ontology as "an explicit specification of a conceptualization" has become the most widely cited in the field of artificial intelligence, other authors have sought to highlight additional important aspects of ontology. First, an ontology is "a formal, explicit description of a shared conceptualization" (Studer et al. 1998). It is an abstract representation of a phenomenon or a part of the world, for which appropriate concepts have been identified (i.e. conceptualization). Definitions of concepts and constraints related to their use should be defined declaratively (i.e. explicitly). In addition, an ontology can be translated into an interpretable language by a (i.e. formal) machine. Finally, an ontology captures consensus knowledge, which is not reserved for the comprehension of a few individuals, but shared by a group or community. The articulation of the ontology with the digital artifact design model needs to be taken into account from these perspectives: "an ontology is a hierarchically structured set of terms for describing a domain that can be used as the skeletal foundation for a knowledge base" (Swartout et al. 1997) and "an ontology provides the means for describing explicitly the conceptualization behind the knowledge base" (Bernaras et al. 1996). An ontology must be engineered based on its intended purpose and use, as well as the services it will ultimately provide.

Mizoguchi (1998) states that the goal of ontological engineering is to specify a conceptualization of the domain of didactic contextualization shared by the members of TEEC (level 1), to formalize it (level 2) and finally to make it operational (level 3), by the implementation of different tools. Thus, in the TEEC project, the purpose of the context ontology was the same as stated by Bernaras et al. (1996), Swartout et al. (1997), then Mizoguchi (1998), namely to conceptualize the skeleton of the context-aware tools' knowledge bases, specifically the Mazcalc skeleton.

As part of the TEEC project, what is the context that we need to model? Bazire and Brézillon (2005) argue that the "context acts like a set of constraints that influence the behavior of a system embedded in a given task". In education, these constraints can take various forms: epistemological, socio-economical or didactic. As an etymological term, "context" derives from the Latin words "contextus", meaning a connection of words or coherence, and "contextere", meaning to weave together. Ontologically, a context is a set of things that are structurally or functionally linked together around a thing in the center. We define the external context we wish to model using the definition by Abowd et al. (1999): context is all the information that can be used to describe a situation, a person, a place and an object. The term "context-aware" refers to a "system (that) uses context to provide relevant information and/or services to the user, where relevancy depends on the user's task". In agreement with Zimmermann et al. (2007), we consider the context along five dimensions: identity, time, space (place), activity and relationship. With the ambition to parameterize context, the idea of modeling it for teaching and learning purposes emerged and led to the development of the CLASH model. Using this model (Forissier et al. 2013), two students living in two distinct contexts have conceptions of the same object that are constructed according to their respective contexts. Therefore, their conceptions of the same object are different. Modeling of the external context of a student can identify context gaps regarding a given object and predict learning themes or specific parameters on which context effects can emerge. This prediction allows learning scenarios to be designed and developed based on context effects.

4.3. The research focuses

The TEEC project focuses on bringing together two different research approaches in a domain where the intellectual traditions are quite different: the French and the Quebecois. Apart from differences in habits and methodological references, the particularity of education research in these contexts is evident in the didactics approach in France and the educational technology approach in Quebec. The TEEC project has been an opportunity to bridge the two approaches, enabling collaboration between seven laboratories in three geographical locations:

– CRREF (*Centre de Recherche et de Ressources en Éducation et Formation*, Université des Antilles), Guadeloupe, overseas France;

– EDA (*Laboratoire de recherches en Éducation*, Université Paris Descartes), Paris, France;

– ESIEA (Engineering School of the Numeric World), Paris, France;

– I3 (*Institut Interdisciplinaire de l'Innovation*, Telecom ParisTech), Paris, France;

– LICEF (*Laboratoire en informatique cognitive et environnement de formation*, TELUQ) Quebec and Montreal, Quebec, Canada;

– CRIA (*Centre de Recherche en Intelligence Artificielle*, (formerly GDAC, *Gestion, Diffusion et Acquisition des Connaissances*), Montreal, Quebec, Canada;

– EREST (*Equipe de Recherche en Education Scientifique et Technologique*), UQAM, Montreal, Quebec, Canada.

Quebec and France tend to ignore the context too often in their teaching practices. The following three cases have been identified: (1) the context is ignored as part of the study of sciences, (2) the immediate context is taken for granted and generalized, (3) the context is recognized as an essential component of science learning and its variation is explored. TEEC aims to make students aware of the importance of context in observation as well as confronting multiple contexts in identifying what constitutes the concept and the example. This general hypothesis has been formulated and adopted: a pedagogical scenario designed to allow context clashes between two different contexts, with the students working on shared themes, produces an awareness from the students of their own representations built up over their lives. A learning design is based on measuring the difference between two contexts using a scientific instrument to determine its capacity for producing the desired effect. The predicted context effect is determined by analyzing the traces of interaction between the students. Ultimately, the goal is to produce a teaching model based on science, duly validated and equipped with digital tools for its implementation. The general hypothesis can be divided into three specific hypotheses.

– Pedagogical scenarios constructed according to the gaps between two external contexts allow a conceptual change in learners' minds.

– By modeling context based on context parameters' characterization, the gap between two contexts can be calculated.

– Validation of the prediction of context effects through context modeling and calculation will take place through the implementation of learning scenarios and by assessing students' conceptions throughout the experiment.

The TEEC project started in 2016 with the following initial objectives.

– Build and test a teaching model based on the emergence of context effects in multiple domains and at various academic levels by setting up multiple pedagogical experiments involving teachers and learners from distant contexts. Create pedagogical scenarios adapted for distant collaboration and with local relevance. Collect data from the experiments to identify and analyze context effects, but also to analyze the learning and teaching.

– Characterize context effects phenomena occurring in the pedagogical experiments using a variety of cross-methodologies: primarily interactional analyses of virtual exchanges (educative platforms, video conferencing, text messages, etc.).

– Develop a digital tool that can model contexts, calculate the context gap between them, and predict the emergence of context effects (Mazcalc).

– Develop digital tools for the design and development of the context-aware intelligent tutoring system (CAITS).

The organization of the numeric tools in the TEEC project as well as their interaction is shown clearly in Figure 4.2. The three context-related components, authoring system, CAITS and Mazcalc are dependent. Mazcalc retrieves information about the learning object from the context pool to provide data regarding the similarities and gaps between the learning object in the two distinct contexts.

The authoring system consists of a number of authoring services (CEM, CAS-Edit, Cas-Viz and CAS-Sim). It retrieves information from the context pool about the teaching context to create the learning design. Furthermore, it retrieves data from Mazcalc to provide advice to the pedagogical designer regarding the relevance of the context effect during pedagogical conception.

CAITS contains classical ITS (intelligent tutoring system) modules: domain, learner and tutor (Nkambou et al. 2010). It is intended to guide learners through their learning, but in a context-sensitive manner, for which it is also interacting with the learner part of the context pool. It diagnoses the learning difficulties encountered by learners and provides adaptive guidance to support learning. It can also provide guidance through the emergence of context effect.

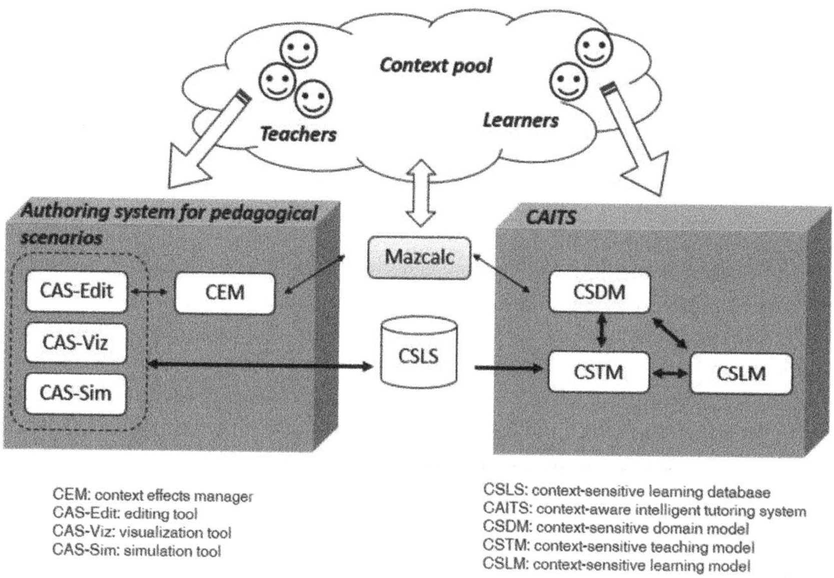

Figure 4.2. *Numeric tools architecture of the TEEC project (from Forissier et al. 2013)*

4.4. Methodology

4.4.1. *DBR methodology*

Design-based research (DBR) methodology has proven to be an appropriate methodology for the TEEC project. It is a micro-systemic methodology that supports both the verification of hypotheses and the testing of innovative instrumentation iteratively (Bourdeau 2017). The design component of this methodology provides methods to improve the learning design and to experiment with it (Figure 4.3).

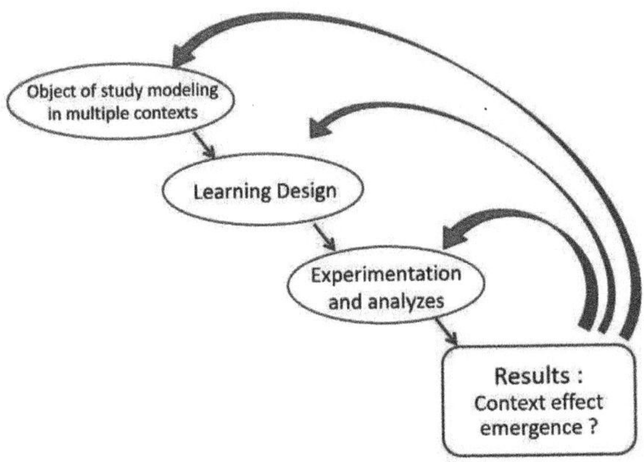

Figure 4.3. *Illustration of the DBR retroactivity in the TEEC project*

In addition, this methodology emphasizes the interaction between research and practice to develop effective learning environments and make use of them as natural laboratories to study learning and teaching (Sandoval and Bell 2004). Five experiments in various domains were conducted for testing the CLASH model in contrasting contexts in Quebec and Guadeloupe, comprising a continental and a tropical context. Details of the experiments are provided in the appendix at the end of the chapter. At the same time, the Mazcalc software was developed to calculate the gap between two external contexts based on various parameters and themes (Anjoù et al. 2017; Psyché et al. 2018).

According to the results of the Mazcalc calculation, learning scenarios can be designed. The learning scenarios must be developed collaboratively between the actors of both contexts. They also need to specify learning activities based on local investigations such as field experiments and lab work, and they must include synchronous and asynchronous exchanges between learners through videoconferencing and educational platforms in order to optimize the emergence of context effects (Forissier et al. 2013). A scenario editor called "Scenarizator" (Miladi et al. 2021) proposes a modeling of learning design specific to context-based collaborative learning that helps teachers and instructional designers in their collaborative design project. Also, efforts toward the design and development of the context-aware intelligent tutoring system and of the authoring system, for designing learning scenarios based on Mazcalc, are underway, along with the experiments.

4.4.2. *Data collection and analysis*

The experiments allow for the collection of various data such as video recordings of synchronous exchanges and of other activities, asynchronous exchanges on the Edmodo platform, pre- and post-test questionnaires, learners' productions, teachers' and learners' interviews and diverse organizational materials (meeting reports, learning scenario drafts, etc.). The collected data, for each experiment are specified in the appendix documents (see section 4.9).

Multiple types of analyses were conducted in diverse fields. Facial emotion recognition (Piot et al. 2019) and interactional analysis (Le Bail et al. 2021a, 2021b) were coupled to characterize context effects. Didactic analysis about the evolution of conceptions, the learning competencies of context effect-based pedagogies, and about the intercultural contribution to the learning process (Chartofylaka et al. 2018, 2019) were carried out to describe the learning events that happened. In addition, analysis in the domain of educational technologies has been conducted and published (Chartofylaka 2020; Stockless et al. 2021). To integrate these analyses and provide recommendations in terms of human–computer interaction, a methodological guide has been developed (Le Bail 2019). A parallel process of designing authentic instruments was also undertaken to facilitate aspects of teaching practices with context effects. This was followed by the analysis of global DBR reflections on the project, discussions with TEEC actors and a literature review. The primary goal is the development and deployment of a context-aware intelligent tutoring system (CAITS). The current effort is focused on the design of a system with two interfaces, one for the author and one for the learner, which both rely on Mazcalc (Forissier et al. 2013).

4.4.3. *TEEC organization*

Figure 4.4 shows the chronology of events during and around the TEEC project, as well as the organization of the various types of work and their interaction. The upper block describes the pedagogical experiments and their analyses, while the lower block shows the organization in the development of the various versions of the digital tools.

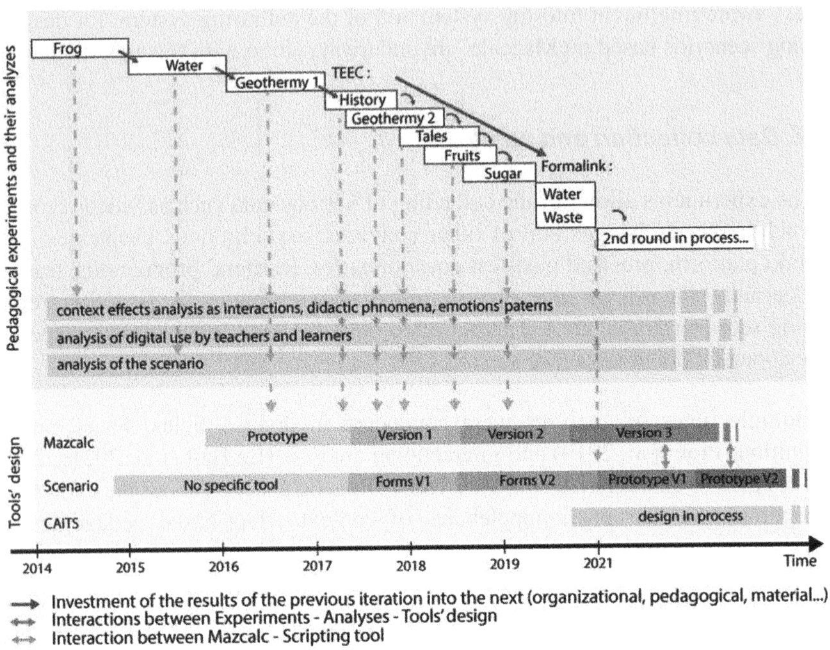

Figure 4.4. *DBR organization of the research activities through iterations. For a color version of this figure, see www.iste.co.uk/brezillon/context.zip*

The simultaneous nature of the various works carried out and in progress makes their schematic representation complex. The main interactions have been represented by arrows but, in general, each type of work constitutes one component of the project and must be considered for the progress and understanding of the research. Each part has feedback that is important for the continuity of the work and that is used for the improvement of the iterations, whether they are experimental, analytical or focused on the development of numerical tools. The retroactive dimension has deliberately not been represented in the diagram in order not to overload it. Note that the implementation of each phase, whether it is experimental, data collection or analysis, the development of tools or the formulation of research hypotheses, implies that observations, data, or results from previous phases must be enhanced and integrated into the realization of the next phase.

4.5. Results and findings

4.5.1. *Context effects identification and specification*

The interactional analyses merged with the facial emotion recognition carried out on the five experiments enables the specification of the modalities of the context effect emergence, the way it manifests itself and its consequences on the interaction.

The interactional analyses are based on theories from the field of computer-supported collaborative learning (CSCL). They have associated the context effect to a phenomenon called "grounding" (Clark and Schaefer 1989; Baker et al. 1999). "Grounding" is the specific process in which learners attain and maintain mutual understanding in dialogue. By this process, learners can develop shared conceptions during an interaction. Due to the initial divergence in learners' conceptions, the grounding is interrupted by a lack of mutual understanding. The interactional sequence of the context effect takes place as follows: (a) questioning (from context 1), (b) a set of unstructured phases of explanation, disagreement, misunderstanding and argumentation (between both contexts) and (c) mutual agreement and understanding from both contexts (Figure 4.5) (Le Bail 2019).

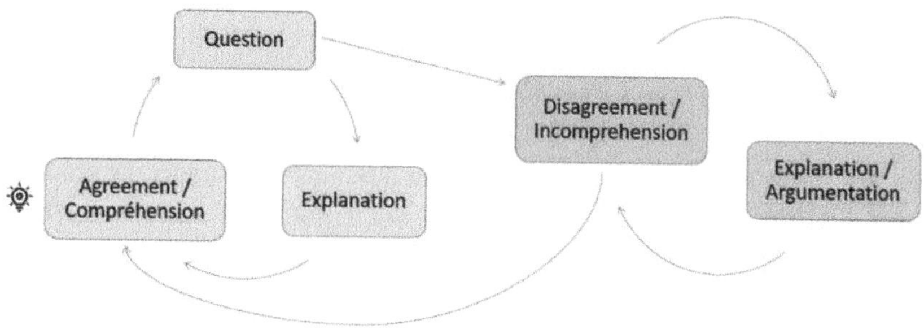

Figure 4.5. *Organization of the interaction during the context effect process, modified from Le Bail (2019). For a color version of this figure, see www.iste.co.uk/brezillon/context.zip*

This context effect emergence model can be illustrated thanks to an example from the experiment on geothermal energy. This example is detailed in the paper by Le Bail et al. (2021a). In this experiment, the learning design was organized so that the students had to investigate the geothermal potential of Guadeloupe and Quebec to make a joint recommendation about the best place to install a data center that would require temperature regulation and a large amount of energy. The example shows how the discussion about the functioning of a heat pump led the students to understand the various sources of heat involved in geothermal energy in Guadeloupe and in Quebec. The exchange took place after a field investigation on both sides. In Guadeloupe, the geothermal power plant uses groundwater at 250°C and 1,000 m deep to produce electricity. In Quebec, the Montreal Biodome, a biological museum that recreates the climate of four ecosystems uses natural groundwater at about 12°C and 20 m deep for temperature regulation.

The transcripts of the interaction above have been summarized and translated from French to English. The assigned code of the speaker G(1 to x) or Q(1 to x) identifies when a student from Guadeloupe or Quebec is speaking, and which one it is. And Tut.G or Tut.Q is used to identify when a tutor (teacher or researcher) from Guadeloupe or Quebec is speaking. The speech turns are identified as N = X.

The sequence used lasted about four minutes and was taken from an exchange during the third session of exchanges between the students of Guadeloupe and Quebec. During this sequence, four students in Quebec and 11 students in Guadeloupe were involved.

N = 001 [Tut.Q]: In Guadeloupe do you understand what a heat pump is and how it works?

N = 004 [G4]: Yes, we understand.

N = 007 [Q1]: Can you explain it?

N = 009 [G4]: ((tilts head forward)).

N = 014 [G11]: Me and G7 we don't understand.

N = 015 [G6]: There is a pump and they said it pumps water under the ground … rather high because with the heat of the sun I think it's warmer at the top ((gestures with hands)).

N = 018 [G11]: ((laughter)).

N = 019 [G6]: Well that's what I understood eh!

N = 022 [Q1]: Yes, that's good! Yes, yes, it makes sense what you say!

N = 025 [G11]: Ok, and it's hot, and it goes through the pipes, and it goes in the house?

N = 027 [G4]: Yes.

N = 028 [G6]: Yes.

N = 029 [Q1]: Yes, that's it, that's what a heat pump does!

N = 036 [G11]: ((to G14)) And in the pump, at first, the water is cold?

N = 037 [G6]: What?

N = 038 [G11]: ((to G14)) At first, the water, if there is water in the pump, is cold and then they heat it?

N = 059 [G11]: Uh …

N = 060 [G4]: Yes.

N = 061 [G6]: No.

N = 062 [G7]: No, no!

N = 063 [G6]: No because we do not use geothermal energy to heat houses like you do … here, it is a big station which aims to generate electricity.

N = 066 [G4]: For everyone.

N = 067 [G11]: For everyone, and then it distributes electricity to each house ((gestures with the hands)).

N = 070 [Q1]: ((to Tut.Q)) OK, so, there is no heat pump. […]

N = 073 [G4]: When we extract water from the borehole after it goes directly into a kind of thing, and after it is mixed with something.

N = 074 [G6]: Yes, because it is steam

N = 076 [Q2]: So, you take the steam directly?

N = 077 [G3]: We pump it, we pump it!

N = 078 [G6]: ((to her group)) Because of the pressure of the steam, no?

N = 079 [G7]: ((to his group)) The water rises directly to our fingers?

N = 080 [Q2]: ((gestures with the hands)) Uh … they take the steam directly?

N = 081 [G3]: We pump it!

N = 082 [G4]: We pump it!

N = 083 [G11]: There is a pump! ((gestures with the hands)).

N = 084 [G4]: There is pressure!

N = 085 [Q2]: ((to themselves)) They take the steam directly?

N = 086 [Q1]: Yeah yes! Because the water is already hot!

N = 087 [Q3]: Basically, the big difference between you and us is, for us, the heat is in the ground, for you, the heat is in the water.

N = 089 [Q1]: Yes, for us, it is in the ground, for you, it is in the water.

N = 090 [G7]: That is!

N = 091 [G11]: That's right!

N = 092 [Q3]: That's right?

N = 093 [G4]: Yes, that's correct!

N = 094 [Q2]: We are geniuses, congratulations to us ((applause)).

This example summarizes the interaction sequence and the process of grounding which occurred in the exchange. The evolution of knowledge is supported by the interactive processes and by the reference to the local context. The discussion first concerns the object of study (heat pump functioning) and does not refer to the contexts. Following that, the discussion concerns the comparison of the two

Geothermal contexts leads to misunderstanding on both sides. The resolution of the conflict relies on explanations from both sides and with the expression of common agreement and understanding of the gap between the two contexts (heat in the ground vs. heat in the water) (Le Bail et al. 2021b). In this example, the activation of the context during explanations allows the mutual understanding.

Facial recognition analyses have been realized by emotional computing specialists Gonzalez and Prevost (Piot et al. 2019; Gonzalez et al. 2021). The analyses of learners' faces during the interactions are based on the six emotions of Ekman (1999). For the geothermal experiment example detailed above, facial recognition analyses were performed. Figure 4.6 shows the intensity of detected joy among the Guadeloupe group on a timeline corresponding to the timeline of verbal interactions in the example presented. The height of the bars represents the number of students for whom the emotion of joy was detected. We can see that joy is mostly detected at the end of the interaction where the bars are the highest.

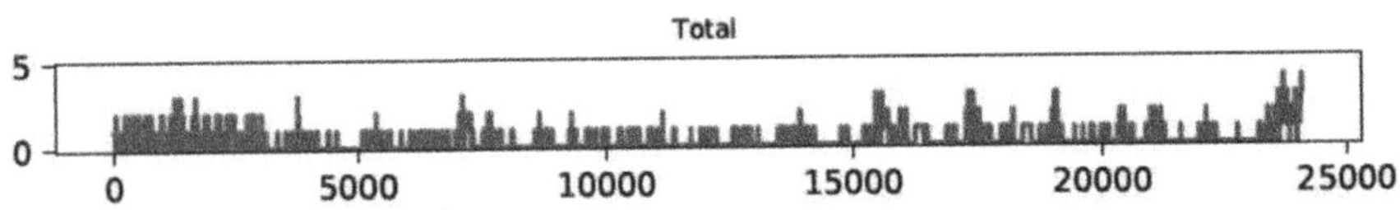

Figure 4.6. *Timeline of the joy detected in facial recognition analyses during the geothermal experiment example (Forissier 2019)*

The superposition of the two types of analyses (the verbal and facial interactions) revealed a match between the emotional moment of joy and the mutual agreement, allowing for a qualification of the context effect according to these two dimensions. This provides a validation of the hypothesis according to which the automatic analysis of emotions can be used to help identify "eureka moments" during interaction and can be used to identify context effects moments.

Further analyses on emotional recognition have permitted the characterization of the context effect as the sudden understanding of a concept by the learners associated with a "eureka" effect. It is associated with emotional patterns resulting in a succession of emotional states of frustration (frown), astonishment (interjection expressed with "Ha!") and joy. It goes without saying that these states are to be put into perspective according to the introverted or extroverted nature of the learners.

4.5.2. *Using the digital technologies*

As mentioned, synchronous and asynchronous support channels were used in the framework of the TEEC project in order to facilitate communication between learners and teachers. A content analysis of their online asynchronous discussions was used to assess the type of information exchanged while using the Edmodo platform[2]. Messages retrieved from this learning management platform were characterized as task-related, group-related and off-task topics, following and readapting the coding scheme framework for online discussions proposed by Vuopala et al. (2016). According to Chartofylaka et al. (2019), several differences have been identified between two of the TEEC experiments that include primary school classes. These differences refer to the different behaviors of learners and teachers when using the platform, the types of messages or media exchanged, to name a few. It is worth mentioning that the richness of the exchanges between participants also depends on many additional elements rather than the motivation and the engagement of participants. Nevertheless, three main points influencing the asynchronous interactions between the participants have been specified: (a) the level of expertise of children and teachers, (b) the technological resources available for use in every school and (c) the Internet connection in classrooms.

Effectively, the Edmodo platform, the digital asynchronous tool selected for the TEEC experiments in primary schools, has proven to be a secure digital environment for young participants to contribute at their own pace to the discussion with no time constraints, as opposed to videoconferences. Individually or in small work groups, it seems that the incorporation of educational social networks can enhance the communication of participants outside the planned synchronous exchanges and in which normally introverted learners are less involved. Findings suggest that the use of Edmodo as an asynchronous tool for collaboration between participants from different contexts should be supported by prior training, coaching and mentoring of the students and the teaching staff.

In essence, analysis of the semi-structured interviews conducted with teachers shows how digital technology can help learner motivation, the development of relationships between actors and the development of learners' skills (Chartofylaka 2020). From a learning point of view, teachers emphasize the contribution of this type of project to the development of students' communication skills. From a technical point of view, teachers gave feedback on their experiences while using the tools, the materials implemented and the general organization of the TEEC project. It has been noted that such projects create additional workload for teachers and this issue should be tackled in future investigations. The learning scenario must be designed in such a way that it supports explicit discussion about contexts

---

2 www.edmodo.com.

(differences and similarities). In addition, a shared work task with a common product is an essential condition for collaboration between learners from diverse contexts supporting a learning potential which is related to grounding and context effects. In fact, this aspect could be managed by researchers who collaborate with the teaching staff to design appropriate common tasks and activities in advance that would foster the emergence of confrontation of contexts during interactions (Le Bail et al. 2021a).

4.5.3. *Learning as an evolution of mental representations*

The analyses resulting from the pre- and post-test questionnaires confirm the conclusions of previous experiments (Forissier 2015; Anjou 2018), which suggested that the design of a pedagogical scenario based on context effects can enhance learning and enrich learners' conceptions (Chartofylaka et al. 2018). Studying the same object in different contexts seems relevant and could enhance learners' knowledge of the given object. Learners' conceptions move from acontextual, spontaneous knowledge (related only to a highly generalized cultural environment) to contextual knowledge (only linked to their own context and to the study of the object in class), bicontextual knowledge (linked to their context and that of the other context, or their context and the very generalized and common cultural environment) or even multicontextual knowledge (linked to the generalized cultural environment, their own environment and that of students in the other territory). This approach allows students to identify what is fully relevant to the concept under study while, at the same time, perceiving how this object adopts specific characteristics in relation to the context in which it is studied. This approach also promotes the evolution of conceptions from spontaneous to expert. According to the initial postulate of the research, the confrontation of contexts cannot be achieved in an "artificial" way (by the teacher presenting different contexts). It is anchored in the contrasting realities of the students, thereby becoming part of an investigative approach that is particularly motivating, where all perspectives are taken into account. This has been demonstrated with analyses of questionnaires issued during experimentation, corroborating the initial hypothesis of the effectiveness of a context effect-based approach in other domains (literature, history) rather than those of natural sciences (Chartofylaka et al. 2018).

According to authors writing about the learning process, context effect-based pedagogical models can be used to develop transversal competencies that are comparable to intercultural competency models (Bartel-Radic 2009; Stier 2009), sustainable development competency models (Tilbury and Mulà 2009; UNECE 2012), digital competency models (Stockless 2018; Stockless et al. 2021) and the

21st century competency model (Romero 2017). Although the competencies are not yet evaluated in the experimentats, the role of context in developing those competencies might be crucial. In future experiments, it would be interesting to evaluate learning by taking inspiration from competency-based models.

4.5.4. *The development of digital tools*

4.5.4.1. *Mazcalc*

Mazcalc, the context calculator, is the first digital tool created for the TEEC project. Although its objectives and uses were well understood by the community of actors involved in the project, the terms used to define it were not in agreement. Therefore, a precise definition, a vocabulary and an ontology were created. In parallel, an Excel prototype was developed and used during the project's experiments. Preliminary results related to this first advance in the tool's development were published (Anjou et al. 2017; Psyché et al. 2018). The tool available during implementation was an Excel prototype and did not function with the other numerical tools designed to assist the pedagogical scenario specification and tutoring system. Thus, it may have been perceived as abstract by the teachers and they may not have understood the link with the concrete learning and teaching activity very well. This observation shows that the development of an intelligent authoring system, including contextual parameters related to the object studies and to the didactic situation, would facilitate the design of learning scenarios based on context effects. Also, it would improve the collaboration between the teachers of the two contexts, and between the teachers and the researchers. The API (application programming interface) was implemented using Visual Studio and based on the context of the geothermal energy pedagogical experiment. It is very well detailed in Fennani's thesis (2020). Some examples of the interfaces will be presented.

The different interfaces of the application enable the creation of a model relative to a study object, for example geothermal energy, and a textual description for this object (Figure 4.7).

Another interface allows us to create the parameters that will describe the object of study inside the model previously created. For each parameter, it is necessary to define the attributes such as the nature of the parameter (ordered or enumerated, qualitative or quantitative) (Figure 4.8). There can be a drop-down list, a value interval or a rule that will apply for the calculation of deviation when the object models have been implemented for several contexts. It is also possible to create a hierarchy in the definition of parameters. Several interfaces are therefore available for the creation and description of parameters.

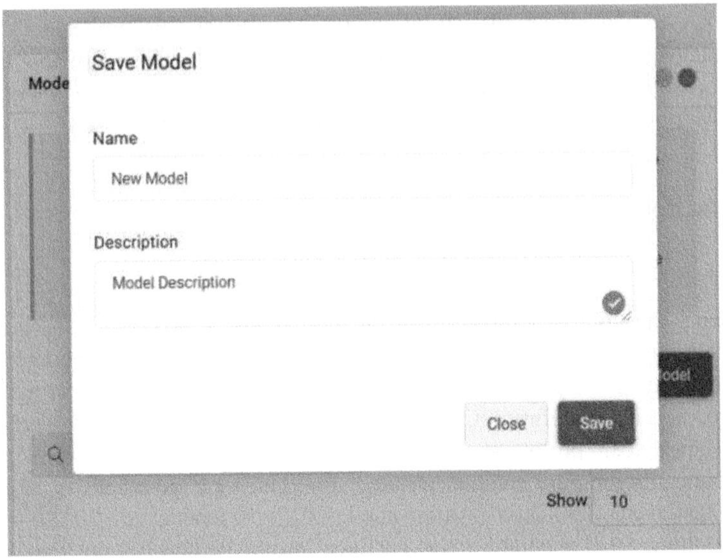

Figure 4.7. *Interface for defining a new study object (Fennani 2020). For a color version of this figure, see www.iste.co.uk/brezillon/context.zip*

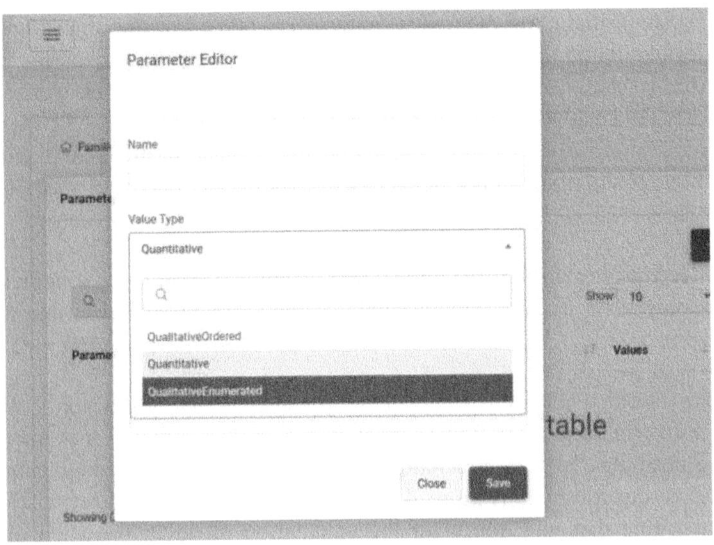

Figure 4.8. *Interface for defining the parameters of the study object (Fennani 2020). For a color version of this figure, see www.iste.co.uk/brezillon/context.zip*

Then, it is possible to implement an existing object model for a given context, for example geothermal energy in Quebec. For each selected parameter, the user must specify the value of the parameter, the degree of confidence to be given to this value and the source of the value. The user can also select only the parameters that are relevant for the implemented context (Figure 4.9).

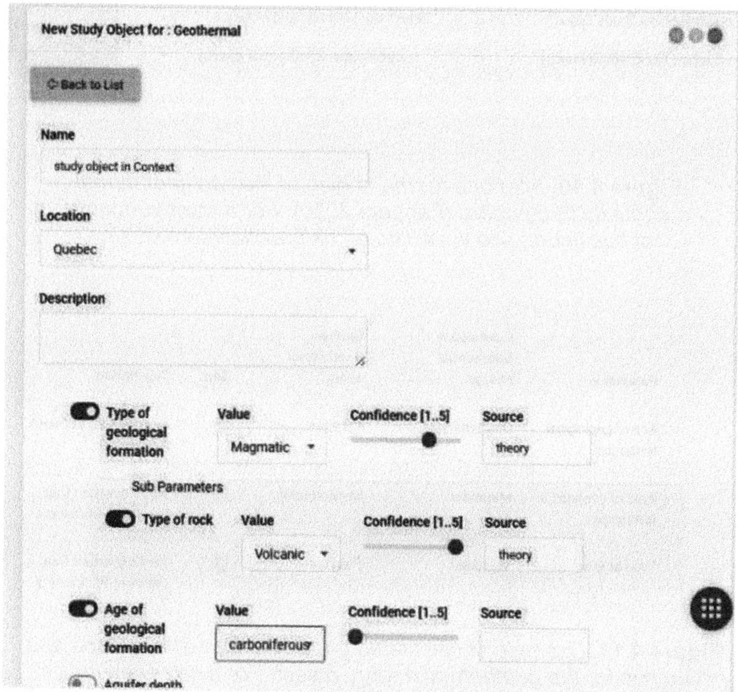

Figure 4.9. *Interface for the instantiation of a context in a study object model (Fennani 2020). For a color version of this figure, see www.iste.co.uk/brezillon/context.zip*

If at least two contexts of the same object of study are defined, it is possible to choose two of them to compare them, as shown in Figure 4.10.

Once the object models have been created, the parameters defined and organized into families and hierarchically ordered, the contexts have been instantiated, the Mazcalc is able to provide a normalized value of the difference between the two selected contexts for each parameter and a description of this value. These values can also be visualized in the form of different types of graphs (Figure 4.11).

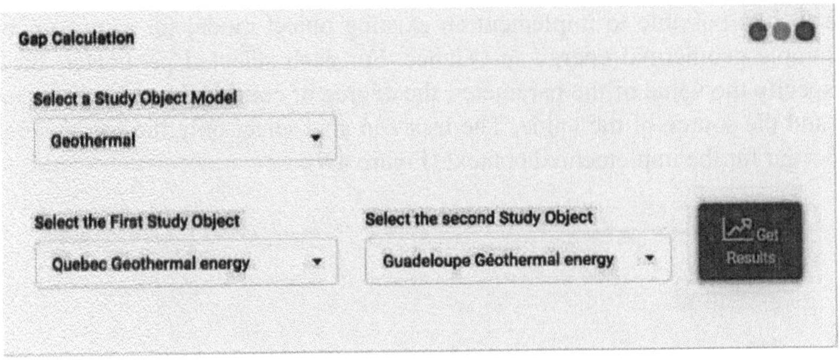

Figure 4.10. *Interface for the choice of the two geothermal contexts to compare (Fennani 2020). For a color version of this figure, see www.iste.co.uk/brezillon/context.zip*

Family	Parameter	Guadeloupe Géothermal energy	Quebec Geothermal energy	Gap	Description	Family Gap
General geology	Age of geological formation	Quaternary	Permian	[illegible]	Average Gap Between Parameters	[illegible]
	Type of geological formation	Magmatic	Metamorphic	[illegible]	Very Important Gap Between Parameters	
	Type of rock	Volcanic	Foliated	[illegible]	Very Important Gap Between Parameters	

Figure 4.11. *Example of calculated gaps between Guadeloupe and Quebec for the geothermal energy object. For a color version of this figure, see www.iste.co.uk/brezillon/context.zip*

The Mazcalc is composed of three modules. First, the "models" module defines the skeleton of a study object. Second, the "context" module instantiates the contexts from the model. This instantiation allows us to determine the relevant parameters for a given context and their associated values. Finally, the "gap calculation" module calculates the difference between two contexts related to the same object of study according to some rules. These calculation rules are also managed by this module.

The Mazcalc has three major strengths. (1) It is a generic tool that is conceived to allow the modeling of any object of study in various fields. (2) It is also designed to instantiate multiple types of contexts. In our study, we were interested in the geographical context, but Mazcalc can be adapted for other contexts such as the temporal context. In this case, the model can be an object of study at a given time and its instantiation in multiple contexts will be done according to the date of data collection. The deviation calculation would thus allow us to study the evolution of the context over time and identify the relevant static and dynamic parameters of the temporal context. (3) Mazcalc can run via its API version as a stand-alone entity on its own website or can be integrated into other systems that should be context-sensitive. CAITS is a relevant example of such a use case.

4.5.4.2. *Scenarizator and the context effect manager*

During the TEEC project, the experiments completed and the retroactive analysis performed allowed us to highlight the importance and complexity of the learning design needed for context effect-based education. Until now, the work carried out during the TEEC project has given little importance to pedagogical scenarios; however, this aspect is very important and should be further studied. The nature of the pedagogical CLASH model requires collaborative (i.e. between stakeholders from both contexts) and contextual (i.e. based on distinct contexts) scenarios. The scenarios have to be constructed collaboratively between teachers or instructional designers of both contexts and include collaboration between learners. They must also consider several aspects such as the synchronization of activities and resources between both contexts, the technical constraints, the digital technologies for exchange, and the physical and pedagogical contexts. Otherwise, the experiments can be compromised. The analysis of the TEEC experiments has highlighted weaknesses in the organizational process that have impacted experiments and which could have been avoided with a formal and rigorous pedagogical scenario writing process.

To avoid this in future experiments, parallel research projects have been implemented since 2021 with emphasis on the design and development of a pedagogical authoring system. An Excel prototype of this authoring system has been developed and adapted with a software specialized in the creation of prototypes and specifications for websites and applications (Axure software). Using the diagram from Forissier et al. (2013) which has been widely used to describe TEEC's objectives on the creation and development of digital tools, we can see the architecture of a contextual tutoring system with its authoring system. The authoring system block is employed with tools for visualizing, editing and simulating pedagogical scenarios. Note that this authoring system interacts with Mazcalc through the context effect manager (CEM), as shown in Figure 4.12.

The CEM is a rule-based expert system that uses expert knowledge to provide guidance to instructional designers and teachers to assist them in the creation of learning scenarios. It interacts with Mazcalc and Scenarizator. It retrieves information about the differences and similarities between two contextual instantiations of a study object from Mazcalc. It then retrieves the information concerning the pedagogical context (school level of the envisaged pedagogical project, organizational and temporal constraints, etc.). Finally, it translates this information into advice that will be visualized through Scenarizator and that will provide assistance to the scenario creators regarding the relevance and feasibility of the scenario (Figure 4.12).

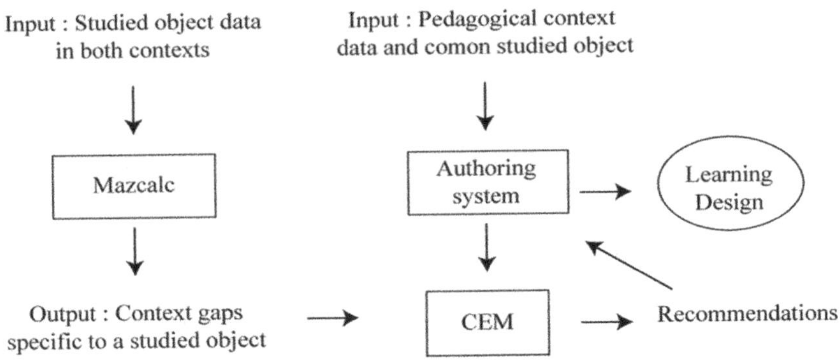

Figure 4.12. *Relationships between Mazcalc, the authoring system and the CEM*

UML specifications of the authoring system have been designed to help a learning designer to describe the learning context and define the academic level, the discipline, the learning objectives and the learning object. The CEM was implemented using CLIPS[3] (C language integrated production system). It is a rule-based expert system with three components: (1) a rule base which is a long-term memory containing rules in the form IF THEN, (2) a fact base which is a short-term memory containing facts and (3) an inference engine that links the rules and the facts to a conclusion. Structured facts have been defined and added to the fact base, and the rules have been coded allowing the inference engine to provide recommendations if the conditions of the rules are fulfilled.

The knowledge model for the authoring system has been created on the GMOT software[5] (an online graphical model editor for the creation of knowledge models, ontologies and learning design). The prototype of the authoring system has been designed on a software that allows the design of dynamic, interactive and functional prototypes and interfaces for websites or applications: Axure software[6] (Anjou et al. 2021). In this prototype, the various interfaces are available and some functionalities have been implemented. For example, the user can search for a learning design by

4 http://www.clipsrules.net.

5 https://gmot.teluq.ca/.

6 The prototype is available at the following link: https://7suikc.axshare.com.

object of study or by context (Figure 4.13). The user can also create a new learning design considering two or more contexts and create the activities, competencies, tools needed and all the necessary information for the implementation of a pedagogical project based on the context effect model (Figures 4.13 and 4.14).

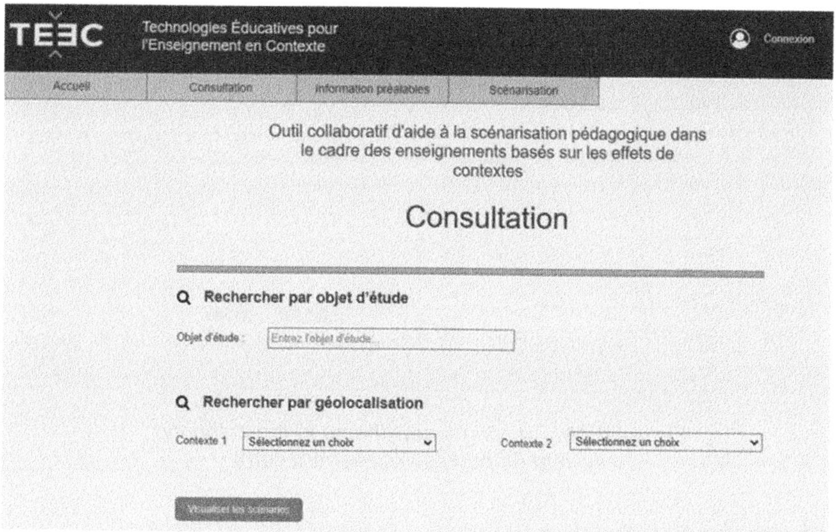

Figure 4.13. *Authoring system search tool (https://7suikc.axshare.com)*

This tool needs to integrate a complex consulting intelligence with multiple parameters regarding the object of study and its respective contexts and also sets up information on the context of the pedagogical project (age of the learners, institutional constraints or technical information on the tools and materials, etc.). The focus should now be on designing the context-aware intelligent tutoring system based on the learners' context and the context of the learning object.

Once the authoring system is developed in the form of a usable digital application, the research data and results of the previous experiments will be implemented within it. It will be then possible to verify the consistency between the CEM recommendations provided and the experimental results. The verification will enable any necessary adjustment of the recommendations. All these data will be stored in a global database of object and context models, learning scenarios and research data. Ultimately, this will lead to a bank of validated learning scenarios that teachers will be able to use to set up pedagogical projects based on the CLASH model.

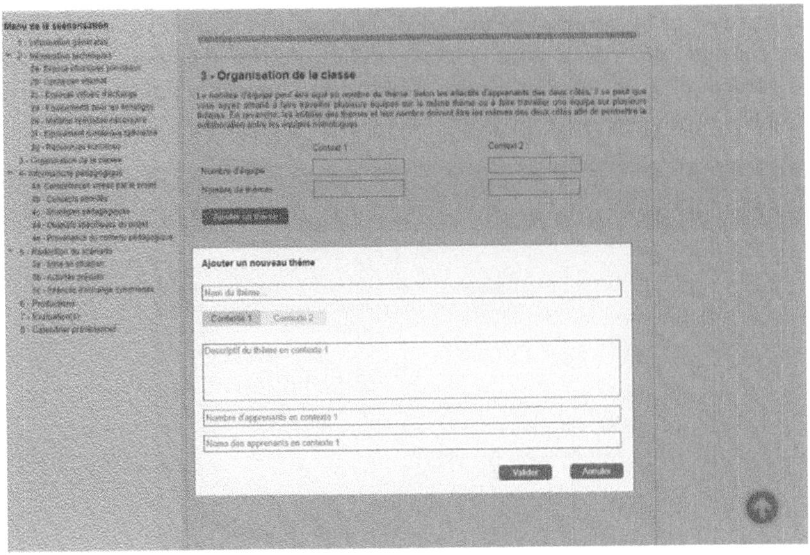

Figure 4.14. *Creating a new theme in the authoring system (https://7suikc.axshare.com)*

4.6. Discussion and interpretation

4.6.1. *Context effect and affective dimension: learning with contexts, contexts effect and cognitive conflict*

The literature in the didactic field has already shown how important the students' prior conceptions are in the learning process (Vosniadou 1994). The results obtained after the didactic analyses performed during the TEEC project showed that the contextuality of students' conceptions is a common dimension in learning.

4.6.1.1. *How can we relate this contextuality to the learning process?*

In the analysis of collected drawings from students through the pre-test and post-test questionnaires of multiple experiments of the TEEC project, the relationship between the contextuality of students' drawings and their scientific relevance was investigated. The categorization of responses can be represented on two axes for the scientific and contextual aspects (Figure 4.15). These two dimensions have also been highlighted in the results of other types of statistical analyses (mostly in multiple correspondences analyses) and in analysis grids realized for the interpretation of pre-test and post-test questionnaires. In most cases, during the experiments drawings become both more scientific and more contextual. This leads us to the following question: Does the construction of contextual conceptions lead to the construction of more expert answers, or does the construction of expert and scientific conceptions lead to the acquisition of contextual and multicontextual conceptions?

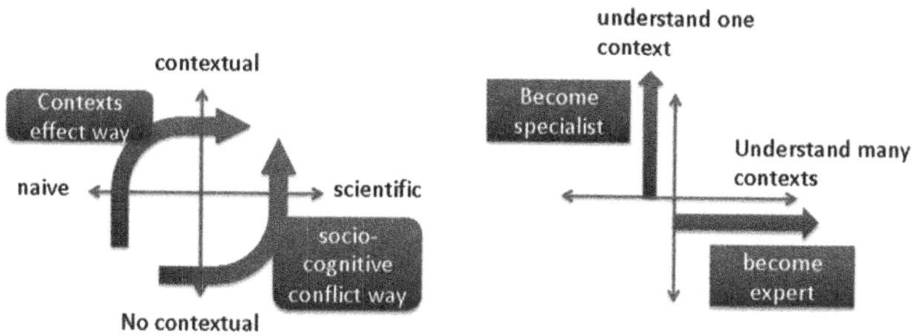

Figure 4.15. *Two representations of the relationship between knowledge and contexts*

We do not have a simple and unique answer; however, this opposition allows us to approach the relationship between context effects and socio-cognitive conflict. Several described context effects correspond to learners taking the path of a socio-cognitive conflict. For example, in an iteration in the field of biology, students working on frog development will spontaneously go to a pond and find tadpoles to observe them and study them in the lab. However, the *Eleutherodactylus* frogs of Guadeloupe are tree frogs and do not have a free-swimming tadpole stage. Students will find *Bufo marinus* tadpoles and later find out what species they belong to. The students came up with this activity because their initial conceptions of the frog, which can be understood as scientifically valid (frogs lay tadpoles in ponds) were not contextually valid. Through investigation, this conception will be confronted by the context which will trigger a socio-cognitive conflict. There is no confrontation here between two particular contexts but between a general idea and a particular context. When students in the nutrition group of the same iteration are surprised and amused to hear that the bullfrog (Quebec) eats mice and small animals, the path taken is not the same. In the investigation, they constructed a contextualized conception of how a frog could feed by observing an *Eleutherodactylus* eating mosquitoes or small butterflies, and they generalized it. They then discovered the contextuality of their conceptions by confronting another context. This does not invalidate their own observations but supplements them using the other context with diversity, variations and other possibilities.

To understand this, we must not place the contextual in opposition to the general, but consider that in the case of naturalistic knowledge, the particular example is not opposed to a general rule or model. A set of individuals' similarities and differences constitute the diversity on which naturalistic scientific expertise is based.

4.6.1.2. *Is the context effect a socio-cognitive conflict?*

What is a socio-cognitive conflict? The definition proposed by Nizet and Bourgeois (1999) is: "The entry into conflict of a given cognitive structure with incompatible information and the resulting cognitive disturbance will engage the subject in the search for a new equilibrium, which will lead, if necessary, to the elaboration of a new structure compatible with the disturbing information". Our view agrees with theirs rather than with Festinger's: "the abandonment of a point of view contradictory to that of a fellow student or an action" (Festinger 1957). The main difference lies in the fact that the socio-cognitive conflict presupposes two contradictory explanatory models, one of which is preferable to the other. In the case of the contexts effect, two contexts shed a different and complementary light on the same complex concept.

How can we differentiate a context effect from a socio-cognitive conflict? If we consider Figure 4.15, we note that one of the limitations of the left part of the figure is the fact that the multicontextual nature of conceptions does not appear, although it is fundamental in contextual learning. Two processes coexist in the process of contextual learning. The first, constructed during the context-based investigation, is the one that results in knowing one's context, potentially to the point of becoming a specialist (comparing with other students from the other context). The second consists of building a multi-contextual conception that we call expert through collaboration with the counterpart group.

With analysis, it is possible to redefine the concept of context effects. It was first defined as a difference linked to the contexts between two conceptions, then as a clash between two internal contexts (Forissier et al. 2013). These psycho-didactic descriptions can be supplemented. The effect of contexts is now defined as a sequence of interactions whose phases have been described: questioning phase (from context 1), set of unstructured phases of explanation, disagreement, misunderstanding and argumentation (between both contexts) and mutual agreement phase (Figure 4.5). The socio-affective dimension of the context effect is also understood through the work of Gonzalez and Prévost around the "eureka effect" which is a moment of change in emotional state observed during lessons through the effects of contexts. We can therefore consider the effect of contexts as a cognitive event caused by the confrontation of two internal contexts, constructed during a collaborative sequence involving learners from two different external contexts and causing changes to individual and collective emotional states. This additive definition leads us to rethink the typology of the context effect in a multidimensional way. The context effect is not only an effect on learning due to the context but exists through the confrontation of investigations carried out in two different contexts. From the point of view of a learning scenario, it is a learning event.

4.6.2. *Digital education and context*

Table 4.1 presents a parallel between the identified principles that can increase the chances of the emergence of a context effect, the learning strategies developed in the pedagogical scenario process and the educational technologies developed for the teaching.

Context effect emergence	Scenario adaptation	Digital technologies
Very different external contexts	Studied object significant for students (from their local context)	External context calculator
	A common pedagogical scenario with a common problem	Videoconferencing moments between the two groups classes
Collaboration between students not necessarily of the same age or school level	Small teams for better privacy	Videoconferencing spaces for small teams' exchanges
	Allow the students to present themselves and meet individually	Chat, forum, access to individual communication solutions
	On field investigations	Tools for in-field data collection (photo, videos, etc.)
Build contextual mental representations	Having a specific methodology	Digital platform for groups
	Writing an answer for each context	Digital platform for teams
	Multiply the moments of collaboration	Present the investigations to the homologue team
Confront mental representations	Alternate between moments of work in context and synchronous and asynchronous exchanges	Digital workbook and weekly feedback on impressions
	Synchronize the presentation of project conclusions	Videoconferences to summarize the project
Build expert mental representations together	Metacognitive phase and clarification of each other's representations	Synchronous synthesis led by both teachers

Table 4.1. *Choice and modality proposed to build teaching based on context effects*

Context-based education cannot exist without the remote learning enabled by digital tools. We can see the importance of the external contexts gap which is manifested in the current examples by geographical distance or territorial discontinuity. The conditions of implementation are therefore quite complex because they must consider the constraints of educational institutions and the digital divide of the territories. In the case of environmental issues, the ability of digital tools to overcome distances can be used to discover the diversity of contexts and the richness of bio- and geodiversity.

4.6.3. *Mazcalc needs to interact with the scripting tool*

An operational version of Mazcalc is being developed as part of the TEEC project. The effective development of a context-aware intelligent tutoring system (CAITS) is now possible. To design effective learning environments or mentoring systems, the architecture of a context-aware system allows the author to provide the system with a learning scenario. In this authoring system, Mazcalc is a key service, used not only to estimate the context gap and therefore the probability of a context effect but also to highlight the context parameters involved. The author could then use this information to author or revise the learning scenario and can be assisted in this adaptation task using scenario management tools. It would thus be possible to play iteratively with the parameters of external contexts provided by Mazcalc and adjust the scenario accordingly. The resulting scenario is stored in a database.

A scenario editor software tool adapted to this type of teaching was designed as part of the TEEC project and named Scenarizator. It is a specification tool that requires the user (author) to specify actors, activities and resources in the scenario. It is instrumental for both the organization of intercultural collaborations and guidance and support for teachers. The different iterations have shown that for practicing teachers, using Mazcalc and the scripting tool is seen as additional workload. It therefore appears necessary to think about how to bring these two objects together. This question fits into a wider context. Teachers rely on their practical epistemic knowledge (Brousseau 1998) or personal content knowledge (Shulman 1986) to design their teaching. One of the challenges is to be able to show the value of this guidance which cannot be reduced to a question of ergonomics.

4.7. Conclusion and related work

The inception of the TEEC project started in 2014 with three pre-experiments in biology, water science and geology. TEEC became official and structured with a grant and a research program designed in 2016. This was done with a team of over 15 researchers in three research groups, with a team of teachers and pedagogical counselors and master's and doctoral students. In addition, we cannot forget the community, school and especially students who took part in our experiments. The benefits for research and education are significant. For now, the results of works carried out in the TEEC project have provided a complete characterization of what a context effect is and of how it arises. Context effects can be identified as a particular interactional and emotional pattern occurring when the relation to context is explicitly formulated in the interaction. The learning design is a complex task that needs to be better framed to provide for real collaboration between students. In addition to the modeling of learners' external context related to the object of study, the pedagogical context (including information related to the educational institution, the resources available or the local instructional culture) could be considered using a new parameter embedded in the authoring system. This would probably improve the global organization of experiments and their progress. The synchronous and asynchronous exchanges platform – that is, a videoconference platform or a learning management system – has proven to provide a good opportunity for collaboration and should be made more accessible to the learners and teachers. Finally, the learning design experiments did not focus on learners' evaluations. Future work could help teachers and instructional designers to reach a common agreement, allowing the collection of data aimed at more precisely identifying the acquired knowledge and the developed competencies. The results could be included in CAITS and improve its intelligence.

In a complex setting with pedagogical innovation, a design-based research (DBR) methodology provides a rigorous guideline to implement our experiments in a real context. The DBR methodology has been explicitly formulated in the TEEC project, and given the systemic and holistic nature of this methodology, the results of previous experiments have been used to improve future experiments. Also, the results lead to some improvements and the development of the instructional approach as well as of digital tools and how we use them for research or pedagogical settings. Those tools must be continuously improved and adjusted as a result of feedback collected from the experiments and analyses undertaken. The analysis methodology developed during the TEEC project may also be applied to future experiments and can be improved with the DBR methodology. Our view for the future includes the creation of two banks: (1) a bank of Mazcalc context instances to serve the purposes of teachers and (2) a bank of learning scenarios that teachers can edit or adapt.

The CLASH model taking a central place in the TEEC project has proven to be adaptable to different age groups, educational levels and in multiple disciplines but, until 2020, all experimentation carried out was conducted between the two same geographical areas: Quebec and Guadeloupe. The new hypothesis to be explored relates to the adaptability of our pedagogical model to other geographical areas. With this ambition, a new project aims to experiment with the context effect-based pedagogical model in schools located in different parts of the archipelago of Guadeloupe (Anjou et al. 2019). Whether and how this model could be applied to multiple languages remains to be shown. The idea of adding a third location and therefore adding complexity is an avenue to be pursued.

4.8. Acknowledgment and credits

The TEEC project was made possible by funding from two research agencies: the *Agence Nationale de la Recherche* (ANR) from France and the *Fond de Recherche du Québec, Société et Culture* (FRQSC) from Quebec, with the project number 2017-QF-210862. The TEEC team wishes to give special thanks to researchers, research assistants, collaborators and especially to teachers who were brave enough to cope with the risks associated with this adventure. Finally, the authors would like to thank Sara Amirsardari for the linguistic revision of the chapter.

4.9. Appendices: description of the TEEC experiments

4.9.1. *Historical event/social realities*

For the socio-historical experiment, the cultural traits (common and specific) of Guadeloupe and Quebec have been exploited in a pedagogical scenario developed with the objective of contributing to the construction of robust conceptions on historical social concepts common to both training programs but which are presented in different ways in the academic programs of both territories.

The pedagogical project was organized between October and November 2017 and involved scholar groups from the Université du Québec in Montréal (UQAM) in Canada and from the INSPE of the Universtité des Antilles at Les Abymes in Guadeloupe. The participants were 12 students on the master's program for school teachers with generalist disciplinary training in Guadeloupe and six students in their second year of training in academic and social adaptation (four-year baccalaureate) with an interest in learning and behavioral difficulties in Quebec. The students from both countries had a disciplinary and generalist profile and were in a history training program (antiquity to the present day). The experiments also involved two teachers (one in each context), two researchers (one in each context), two students in Guadeloupe (a doctoral and a post-doctoral student) and one master's student in Quebec. Each group of students was organized into three teams working in a homologous way on three themes: (1) power, monarchy and absolutism; (2) revolution, separation of powers and political regimes; (3) machinization, mechanization, urbanization and industrialization.

The experiments allowed the collection of four filmed sessions in group classes, pre- and post-test questionnaires, one student production (a video clip) and an interview with the teacher from Quebec.

4.9.2. *Geothermal energy*

The geothermal energy experiments involved students from Le Vitrail alternative school (based on an active pedagogy) in Montreal, Quebec and students from Edmond Bambuck secondary school in Le Gosier, Guadeloupe. It also involved 14 students in Guadeloupe and 13 students in Quebec aged 12 to 14. Some students in Quebec abandoned the project after the first session and only a small group of five students continued the project to the end. Three teachers (one in Quebec and two in Guadeloupe), three researchers (two in Guadeloupe and one in Quebec) and a doctoral student in Guadeloupe were also involved.

The experiments were designed based on the following pedagogical scenario:

> The company Google manages a very large and growing data center park. The company tries to do everything possible to ensure that its data centers are 100% powered by renewable energy. The servers have to run 24 hours a day, and the cooling system, which is needed to remove the heat generated by the servers, is very energy intensive. This makes Google data centers' reliance on local power providers very costly for the company. Geothermal energy is a technology that could be used to reduce the energy consumption of data centers. Could Quebec and Guadeloupe be suitable locations for a data center using geothermal energy? What type of technology could be used to do this? Two teams will work together, one in Guadeloupe, the other in Quebec, to study the geothermal potential of the two regions and make a recommendation to Google for the establishment of its data center. Frequent exchanges through technological communication tools will occur between the two groups.

During these experiments, the students were organized in teams on each side, working on the five following themes: (1) the geological context, (2) the geothermal potential, (3) the principle of physics, (4) the technology and (5) the climate.

The experiments allowed the collection of four filmed sessions in group classes (with only one team in Quebec after the first session), pre- and post-test questionnaires about the students' mental representation of geothermal energy, pictures and a presentation from the students after the project who participated and won a national competition, "C Génial".

4.9.3. *Literature*

The two scholar groups that participated in the literature experiment are a CM2 class in Guadeloupe, composed of 20 students aged from 10 to 11 years (10 boys and 10 girls) and a 5th and 6th grade class (16 students in total) in Quebec aged from 10 to 12 years old (eight boys and eight girls). The scholar groups come from the Turenne Thénard school in Le Gosier in Guadeloupe and from the Chanoine-Ferland school in Saint-Fabien-de-Panet in Quebec. The experiment also involved two teachers (one in each context), three researchers and one doctoral student in Guadeloupe and a researcher and a master's student in Quebec. The experiment lasted two months from April to May 2018. It included the collection of six filmed sessions: three videoconferences with the class group lasting from 30 to 45 minutes each (a total of two hours of recording), three videoconferences of the five mirror teams lasting 20 minutes, pre- and post-questionnaires and one teacher interview.

The pedagogical scenario focuses on the study and rewriting of a tale: "Compère Lapin et Compère Zamba" in Guadeloupe and "Rose la tulipe" in Quebec. The tale is an ideal medium for: (1) in literature, addressing notions such as the structure of the tale, the attributes of the characters and the oral tradition; (2) in language mastery, confronting different varieties of French (Quebec and West Indian). It allows students to work on the definition of what a language is and the variety of actualizations of the same language depending on where it is spoken. The tale is also an ideal object to explore the cultural field of a territory. The pedagogical scenario includes different working times: (1) work by the teacher in their class with their students, (2) work by the team in class, (3) exchange times by videoconference with the other class and (4) exchange times by videoconference between the mirror teams of each territory. Five themes were chosen to get the students to work on their respective tales and to get them to exchange: (1) the narrative scheme, (2) the characters, (3) the places, (4) the expressions and the lexicon and (5) the cultural references. The tales selected are two typical tales from each territory. The teachers each proposed their tale with the objective that it be adapted to their students and that it be well known and representative of tales in the territory.

4.9.4. *Sustainable development: sugar*

The "Sugar" experiment involved 9- to 10-year-old students corresponding to the 4th grade in Quebec, and CM1 in Guadeloupe: 26 students from La Mennais school in Quebec and 23 students from the elementary school of Pombiray at Saint-François in Guadeloupe with their respective teacher. The pedagogical project lasted three months from February to April 2018. The first step of the experiment involved a preparation phase in collaboration between the actors on both sides for technical organization and recording settings, for the coherence of the pedagogical content, the learning scenario, the teaching and the sharing of activities, and for synchronization between Guadeloupe and Quebec. This step lasted almost four months. A first videoconference was organized during which every student had the opportunity to present themselves to the other group and to present a short phrase describing the school or the local context (geography, tradition, climate, food, activities, etc.). Then, students from both sides were divided into six teams working on six themes: history/culture, biology, agriculture, economy, health and ESD. A team from each context worked on a specific theme and each team was involved in scientific inquiry in their local context. The inquiry involved formulating questions linked to the theme (history, biology, etc.), either about the cane or the maple depending on the context, to answer those questions throughout the various activities and using the Internet, media and books, and in the end, to produce a poster with a brief presentation.

In Quebec, three activities that related to maple syrup were organized: (1) visiting a sugar shack where the process of collecting the maple water and transforming it into maple syrup was explained. The students were invited to participate so they could understand the process. They also tasted the various maple products (maple sugar, sugar loaf, maple butter, etc.). The maple producer also explained a bit about the history and discovery of the maple and about the biological phenomenon that allows the fluid circulation in the tree; (2) intervention from a nutritionist who talked to the students about food and health related to maple syrup; (3) intervention from a research scientist specializing in the ecology of forests and climate change who presented the impact of climate warming on maple forests and maple production.

In Guadeloupe, two activities linked to sugar cane were organized: (1) there was a visit to a Guadeloupe plantation housing a rum distillery and a creole garden with a small train tour (the Severin domain). The visit involved an the explanation of the rum process from specialists; (2) a visit to two sugar cane plantations (Mahaudière and Néron) with a heritage curator who explained the history of the sugar cane, its importation and its link to slavery and colonization.

During the activities, students could ask the speakers questions. They also had the responsibility of recording videos and taking pictures and notes to share with the students from the other context. The objective of the exchanges was to expose students to the conceptions of students from another context, to provide them with new ideas for questions, to motivate them to find answers and to expose them to global issues through the various themes. The students also had the opportunity to exchange messages, photos and videos via the Edmodo educative platform. At the end of the experiment, they also sent each other gifts (sugar and maple samples).

The experiment allowed the collection of eight filmed videoconference sessions between two group classes (85 minutes) and three between mirror teams (148 minutes), five filmed activities, 75 posts on the Edmodo platform, student productions (10 documents), pre- and post-test questionnaires and two teacher interviews.

4.9.5. *Sustainable development: fruit*

For the "Fruit" experiment in Quebec, the same teacher as for the "Sugar" experiment participated and 26 students from La Mennais school in Mascouche were enrolled. In Guadeloupe, 21 students from the Amede Fengarol school in Capesterre-Belle-Eau took part in the project along with their teacher. The preparation phase started in April 2018 and the pedagogical experiment started in September, lasting five months until January 2019. The goal of this experiment was to focus on sustainable development issues and improve the organization by providing teachers with propositions for activities, questions and content for each theme. This experiment started with a presentation of the project and of what sustainable development means by researchers familiar with sustainable development in Guadeloupe and in Quebec. To stimulate interactions between students during this experiment, the students were asked to answer a common question: "Which one of the two fruits (apple or banana) is the best for sustainable development?". A first videoconference with both groups was set up for the students to present themselves and say a short sentence about the school or local context. As with the Sugar experiment, the classes were organized into six teams investigating six themes regarding the banana in Guadeloupe and the apple in Quebec: biology, agriculture, culture and history, health, environment and economy. A first set of videoconferences between teams was organized for the students to present themselves better and get to know one another. Then, two more sets of videoconferences between teams were set up. The first was for an exchange of questions and the hypothesis, and the second for the presentation of collected information during surveys and visits, as well as for answering questions from the previous videoconference. Then, a global videoconference took place for the classes to present their final production. In Quebec, the visit of an orchard with an apple picking was organized. In Guadeloupe, the students discovered the peculiarities of the banana along with the visit of the CIRAD (Agronomic International Research Center for Development) and of a banana plantation. During the visits in both contexts, experts were solicited to explain the learning object to the students.

The experiment allowed the collection of eight filmed sessions of videoconferencing, two between group classes (170 minutes) and six between mirror teams (440 minutes), 34 posts on the Edmodo platform, student productions

(16 documents including slideshows and video clips), pre- and post-test questionnaires, pictures and a teacher interview.

4.10. References

Abowd, G.D., Dey, A.K., Brown, P.J., Davies, N., Smith, M., Steggles, P. (eds) (1999). Towards a better understanding of context and context-awareness. *International Symposium on Handheld and Ubiquitous Aomputing*, 304–307.

Anciaux, F. (2013). Alternances et mélanges codiques dans les interactions didactiques aux Antilles et en Guyane françaises. PhD Thesis, Université des Antilles-Guyane, Guadeloupe.

Anjou, C. (2018). L'énergie géothermique : représentations, enseignements et contextes aux Antilles et dans la zone caraïbe. PhD Thesis, Université des Antilles, Guadeloupe.

Anjou, C., Forissier, T., Bourdeau, J., Mazabraud, Y., Nkambou, R., Fournier, F. (eds) (2017). Elaborating the context calculator: A design experiment in Geothermy. *International and Interdisciplinary Conference on Modeling and Using Context*, 513–526.

Anjou, C., Forissier, T., Delcroix, A. (2019). FOMALINK: A collaborative pedagogical project in the Caribbean. *Hyper Article en Ligne – Sciences de l'Homme et de la Société*, ID: 10670/1.hfznqj.

Anjou, C., Miladi, F., Psyché, V., Bourdeau, J., Forissier, T., Stockless, A. (2021). Development of an intelligent, context sensitive and collaborative authoring system for context effect-based education. *Modeling and Using Context*, 4 (CONTEXT-21) Special Issue.

Baker, M., Hansen, T., Joiner, R., Traum, D. (1999). The role of grounding in collaborative learning tasks. In *Collaborative Learning: Cognitive and Computational Approaches*, Dillenbourg, P. (ed.). Elsevier, Bingley.

Bartel-Radic, A. (2009). La compétence interculturelle : état de l'art et perspectives. *Management International*, 13(4), 11–26.

Bastien, C. (1998). Contexte et situation. In *Dictionnaire des Sciences Cognitives*, Houdé, O., Kayser, D., Koenig, O., Proust, J., Rastier, F. (eds). PUF, Paris.

Bazire, M. and Brézillon, P. (2005). Understanding context before using it. *Proceedings of the 5th International and Interdisciplinary Conference on Modeling and Using Context*. Paris.

Bernaras, A., Laresgoiti, I., Bartolome, N., Corera, J. (eds) (1996). An ontology for fault diagnosis in electrical networks. *Proceedings of the International Conference on Intelligent System Application to Power Systems*, IEEE, 199–203.

Blanchet, J., Moore, D., Asselah Rahal. S. (2009). *Perspectives pour une didactique des langues contextualisee*. Editions des archives contemporaines, Paris.

Bourdeau, J. (2017). The DBR methodology for the study of context in learning. In *Modeling and Using Context. CONTEXT 2017. Lecture Notes in Computer Science*, Brézillon, P., Turner, R., Penco, C. (eds). Springer, Berlin, Heidelberg.

Brousseau, G. (1998). *Théorie des situations didactiques*. La Pensée Sauvage Éditions, Grenoble.

Brown, A.L. (1992). Design experiments: Theoretical and methodological challenges in creating complex interventions in classroom settings. *Journal of the Learning Sciences*, 2(2), 141–178.

Brown, J.S., Collins, A., Duguid, P. (1989). Situated cognition and the culture of learning. *Educational Researcher*, 18(1), 32–42.

Chartofylaka, L. (2020). Enrichir les pratiques de classe par le numérique : retour des enseignants sur une expérimentation menée en Guadeloupe. *Rencontres Jeunes Chercheurs en EIAH (RJC EIAH 2020)*. Poitiers.

Chartofylaka, L., Anciaux, F., Candau, O.-S., Jeannot-Fourcaud, B., Carignan, I., Saint-Pierre, A. (2018). Il était une fois des contes en contextes: Retour sur une expérimentation au primaire en Guadeloupe et au Québec. *Contextes et didactiques*, 11, 67–108.

Chartofylaka, L., Stockless, A., Fraser, M., Psyché, V., Forissier, T. (2019). Sharing contextual knowledge information via asynchronous distance learning: Insights from a context-based research project in primary schools. *Médiations et médiatisations* (2), 117–130.

Clark, H.H. and Schaefer, E.F. (1989). Contributing to discourse. *Cognitive Science*, 13(2), 259–294.

Delcroix, A., Forissier, T., Anciaux, F. (2013). Vers un cadre d'analyse opérationnel des phénomènes de contextualisation didactique. In *Contextualisation didactiques : approches théoriques*, Anciaux, F., Forissier, T., Prudent, L.F. (eds). L'Harmattan, Paris.

Delcroix, A., Cariou, J.-Y., Ferrière, H., Jeannot-Fourcaud, B. (2015). *Apprentissages, éducation, socialisation et contextualisation didactique : approches plurielles*. L'Harmattan, Paris.

Duru-Bellat, M. and Mingat, A. (1988). Le déroulement de la scolarité au collège : le contexte fait des différences. *Revue française de sociologie*, 649–666.

Ekman, P. (1999). Basic emotions. In *Handbook of Cognition and Emotion*, Dalgleish, T. and Power, M. (eds). John Wiley and Sons, New York.

Fécil, S. (2014). Construire un enseignement en tenant compte des effets de contexte. Document, ESPE de Guadeloupe.

Fennani, W. (2020). Vers des systèmes d'apprentissages sensibles au contexte : développement d'un sous-système de modélisation et d'estimation d'écart de contextes pour la prise en charge des effets de contextes. PhD Thesis, UQAM, Canada.

Festinger, L. (1957). *A Theory of Cognitive Dissonance*. Stanford University Press, Redwood City.

Forissier, T. (2015). Conceptions d'étudiants scientifiques de Guadeloupe sur l'observation de l'orientation de la lune et des saisons climatiques. In *Apprentissages, éducation, socialisation et contextualisation didactique: Approches plurielles*, Delcroix, A., Cariou, J.-Y., Ferriere, H., Jeannot-Fourcaud, B. (eds). L'Harmattan, Paris.

Forissier, T. (2019). Contextualisation et effets de contextes dans l'apprentissage des Sciences. Accreditation to Supervise Researches (HDR), PhD Thesis, Université des Antilles, Guadeloupe.

Forissier, T., Bourdeau, J., Mazabraud, Y., Nkambou, R. (2013). Modeling context effects in science learning: The CLASH model. In *Modeling and Using Context: 8th International and Interdisciplinary Conference, CONTEXT 2013, Annecy, France, October 28–31, 2013, Proceedings*, Brézillon, P., Blackburn, P., Dapoigny, R. (eds). Springer, Berlin, Heidelberg.

Gonzalez, J., Prevost, L., Piot, M., Alabarbe, T. (2021). Analyse des interactions émotionnelles dans les apprentissages collaboratifs. *Proceedings of the 88th Colloque de l'Acfas*, Quebec, Canada.

Gruber, T.R. (1993). A translation approach to portable ontology specifications. *Knowledge Acquisition*, 5(2), 199–220.

Le Bail, C. (2019). Apprentissage basé sur la confrontation à distance de contextes. Analyse des interactions verbales, caractérisation des effets de contexte et recommandations. Postdoctoral Research Report. Paris Télécom et Université Paris V Descartes, France.

Le Bail, C., Baker, M., Détienne, F., Bernard, F.-X., Chartofylaka, L. et Forissier, T. (2021a). Grounding and knowledge elaboration across cultural and geographical contexts: An exploratory case study [Online]. Available at: https://www.sciencedirect.com/science/article/pii/S2210656120301483?dgcid=coauthor.

Le Bail, C., Détienne, F., François-Xavier, B. Baker, M. (2021b). La co-élaboration des connaissances dans l'apprentissage basé sur la confrontation à distance de contextes : analyse de Séquences interactives avec Effet de Contexte (SEC). *Proceedings of the 88th Colloque de l'Acfas*, Quebec, Canada.

McCarthy, J. (1980). Circumscription – a form of non-monotonic reasoning. *Artificial Intelligence*, 13(1–2), 27–39.

Miladi, F., Anjou, C., Psyché, V. (2021). Un outil de scénarisation pédagogique conscient du contexte. *Proceedings of the 88th Colloque de l'Acfas*, Quebec, Canada.

Mizoguchi, R. (eds) (1998). A step towards ontological engineering. *12th National Conference on AI of JSAI*, Tokyo, Japan.

Nizet, J. and Bourgeois, É. (1999). *Apprentissage et formation des adultes*. Presses universitaires de France, Paris.

Nkambou, R., Mizoguchi, R., Bourdeau, J. (2010). *Advances in Intelligent Tutoring Systems*. Springer Science & Business Media, Berlin, Heidelberg.

Pearl, J. (1988). Embracing causality in default reasoning. *Artificial Intelligence*, 35(2), 259–271.

Piot, M., Alabarbe, T., Gonzalez, J., Le Bail, C., Prevost, L., Bourdeau, J., Bernard, F.-X., Baker, M., Détienne, F. (2019). Joint analysis of verbal and nonverbal interactions in collaborative E-learning. *2019 8th International Conference on Affective Computing and Intelligent Interaction Workshops and Demos (ACIIW)*, pp. 1–5.

Psyché, V., Mendes, O., Bourdeau, J. (2003). Apport de l'ingénierie ontologique aux environnements de formation à distance. PhD Thesis, Université du Québec à Montréal.

Psyché, V., Anjou, C., Fennani, W., Bourdeau, J., Forissier, T., Nkambou, R. (eds) (2018). Ontology-based context modelling for designing a context-aware calculator. *ITS 2018 Workshop Proceedings*, Montreal.

Romero, M. (2017). Les compétences pour le XXI e siècle. In *Usages créatifs du numérique pour l'apprentissage au XXIe siècle*, Romero, M., Lille, B., Patiño, A. (eds). PU, Quebec.

Russell, S. and Norvig, P. (2010). *Artificial Intelligence: A Modern Approach*, 3rd edtion. Prentice Hall, Hoboken.

Sandoval, W.A. and Bell, P. (2004). Design-based research methods for studying learning in context: Introduction. *Educational Psychologist*, 39(4), 199–201.

Schwartz, R., Lederman, N., Crawford, B. (2004). Developing views of nature of science in an authentic context: An explicit approach to bridging the gap between nature of science and scientific inquiry [Online]. Available at: https://onlinelibrary.wiley.com/doi/pdfdirect/10.1002/sce.10128.

Shulman, L.S. (1986). Those who understand: Knowledge growth in teaching. *Educational Researcher*, 15(2), 4–14.

Sowa, J.F. (1991). Toward the expressive power of natural language. In *Principles of Semantic Networks*, Borgida, A. (ed.). Elsevier, Amsterdam.

Stier, J. (2009). Internationalisation, intercultural communication and intercultural competence [Online]. Available at: https://immi.se/intercultural/nr11/stier.pdf.

Stockless, A. (2018). Le numérique en éducation : apprendre en ouvrant les murs de la classe. *Médiations et médiatisations*, 1(1), 3–3.

Stockless, A., Forissier, T., Chartofylaka, L., Locret, S., Psyché, V. (2021). Étude des usages du numérique dans les apprentissages basés sur les effets de contextes. *Proceedings of the 88th Colloque de l'Acfas*, Quebec.

Studer, R., Benjamins, V.R., Fensel, D. (1998). Knowledge engineering: Principles and methods. *Data & Knowledge Engineering*, 25(1–2), 161–197.

Swartout, B., Patil, R., Knight, K., Russ, T. (1997). Towards distributed use of large-scale ontologies [Online]. Available at: https://www.aaai.org/Papers/Symposia/Spring/1997/SS-97-06/SS97-06-018.pdf.

Tilbury, D. and Mulà, I. (2009). *Review of Education for Sustainable Development Policies from a Cultural Diversity and Intercultural Dialogue: Gaps and Opportunities for Future Action*. UNESCO, Paris.

UNECE (2012). Learning for the future: Competences in education for sustainable development. Report, UNECE, Geneva.

Van Wissen, A., Kamphorst, B., Van Eijk, R. (2013). A constraint-based approach to context. In *Modeling and Using Context*, Brézillon, P., Blackburn, P., Dapoigny, R. (eds). Springer, Berlin, Heidelberg.

Vosniadou, S. (1994). Capturing and modeling the process of conceptual change. *Learning and Instruction*, 4(1), 45–69.

Vuopala, E., Hyvönen, P., Järvelä, S. (2016). Interaction forms in successful collaborative learning in virtual learning environments. *Active Learning in Higher Education*, 17(1), 25–38.

Wertsch, J.V. (1991). A sociocultural approach to socially shared cognition. In *Perspectives on Socially Shared Cognition*, Resnick, L.B., Levine, J.M., Teasley, S.D. (eds). American Psychological Association, Washington D.C.

Zimmermann, A., Lorenz, A., Oppermann, R. (eds) (2007). An operational definition of context. *Proceedings of the 6th International and Interdisciplinary Conference on Modeling and Using Context*, 558–571, Roskilde.

5

Pragmatic Reasoning in Context: Context-Mediated Behavior

In this chapter, we discuss a pragmatic approach to context-based intelligent reasoning, context-mediated behavior (CMB). This approach has been developed over the past 30+ years in several domains, both for autonomous agents as well as for multiagent systems. CMB explicitly represents contextual information as *contextual schemas* (c-schemas), which represent classes of situations that have implications for how an agent should behave. Contexts are treated as first-class objects that can be reasoned about, compared to one another and to the agent's current situation, and used to guide reasoning and behavior. Contextual knowledge contained in c-schemas includes descriptive knowledge about the situations represented (including context-specific meaning of concepts and predictions about unseen or future features) and prescriptive behavior about how to behave (suggestions for plan-like structures, importance estimates of goals, knowledge about how to detect and handle unanticipated events, and standing orders for context-specific behavioral parameter settings). An agent's current situation is diagnosed to be one or more c-schemas, whose knowledge is merged to guide the agent while in the context. We discuss the approach and trace the origins of CMB and its evolution over time.

5.1. Introduction

Context is critical to any agent that operates in the real world, or indeed, in any open world with much complexity (see Chapters 1 and 3). This is true for both natural and artificial agents. Although dancing is a perfectly appropriate thing to do in the context of a nightclub, it would be seen as bizarre, not to mention dangerous, if someone were dancing in the middle of a busy highway: the action is simply

---

Chapter written by Roy M. TURNER.

inappropriate for the context. Similarly, an autonomous underwater vehicle (AUV) would be acting perfectly rationally to surface to update its position from a GPS satellite when in the open ocean, but would not be if it knew it was in the context of being under ice and tried to surface anyway. Rational behavior is context-dependent. Part of the knowledge that must be taken into consideration includes knowledge of the context.

Artificial intelligence (AI) systems have always incorporated *some* contextual knowledge[1]. Early theorem provers contained axioms related to the context of problem-solving, most often with the context-related aspects buried in the antecedents of implications. Early planning systems (e.g. Fikes and Nilsson 1971) often included *frame axioms* that explicitly stated facts about the domain in which the system was to operate (e.g. "an object's position does not change unless the agent moves the object"), and later planners often incorporated *filter conditions* in the descriptions of their actions that served to check the applicability of an operator against the current context. Similarly, early rule-based expert systems (e.g. Shortliffe 1976) often included a fact in their working memory that labeled the current context with a symbol; rules antecedents could then contain clauses that checked if the context was appropriate for their use, and rules could also change the context by retracting the current context fact and asserting a new one.

However, it was recognized relatively early in the history of AI that this was insufficient. Instead, context came to be seen as something that should be represented explicitly so that it could be reasoned about and help organize the agent's knowledge, as well as allow for different, even contradictory, knowledge to reside comfortably and usefully in the same knowledge base. For example, in the formal side of AI, there was McCarthy's (1993) seminal work on context formalization via the *ist()* predicate and lifting rules and Giunchiglia's (1993) bridge rules. Guha's (1992) *microtheories* were an attempt to partition a reasoner's knowledge into contexts in which the knowledge was useful, as were such approaches in the rule-based expert system community, such as CENTAUR (Aikins 1980), MDX (Chandrasekaran et al. 1979), MDX2 (Sticklen 1987) and SEPT (Brézillon 1990).

Around this time, context was also beginning to be taken into account, at least to some extent in the less formal side of AI, drawing inspiration from cognitive science, as well as the very pragmatic needs of problem-solving in complex worlds. Early work on using scripts for story understanding (Cullingford 1981), for example, allowed a

1 We are not here concerned with the long history of studying context with respect to language, for example in computational linguistics, natural language processing and philosophy. While these fields and their work remain highly relevant to the context community, our focus here is on AI systems that take action instead of or in addition to communicating with others.

program to answer questions about the events in a story that were not provided directly by referring to a script containing stereotypical actions and events that transpire in the story's context. Case-based reasoning (see, for example, Kolodner (1993)) embodied a kind of context representation in the form of records of past problem-solving (cases) that can be retrieved and used to help in a new situation, although context effects here were not separated from problem-solving actions until reuse, if then.

In the mid- to late 1990s, a community of researchers emerged, drawn from across the spectrum of formal, purely pragmatic and cognitive approaches, that focused on investigating context as a phenomenon in its own right and how context affects behavior in natural and artificial systems. This community began to coalesce due to a workshop at the International Joint Conference on AI in 1993 (Brézillon 1993; Brézillon and Abu-Hakima 1995) and led to the creation of the biennial International and Interdisciplinary Conference on Modeling and Using Context (CONTEXT) conference series, whose first meeting was in 1997 (Brézillon and Cavalcanti 1997). This conference series, now 25 years later, continues to be the heart of the community. Another community, this one focused on non-AI context-aware software applications, also arose (e.g. Dey et al. 2001), with connections between the two communities becoming common as time went on.

As detailed in Chapter 1, the CONTEXT community has increasingly come to focus on what may be called *pragmatic* research, as opposed to formal, top-down prescriptions for modeling and using context. This has been key to addressing problems in real-world settings, such as responding to subway incidents (e.g. Brézillon et al. 1997) and representing and mimicking human tactical reasoning (e.g. Gonzalez and Brézillon 2008). These approaches, although informed by both formal and cognitive work in AI, have emphasized problem-solving first and formalism and cognitive plausibility later, and they have taken a much richer view of context than is common in prior work or in the context-aware applications community.

This chapter reports on work in the author's lab on one long-running approach to pragmatic contextual reasoning called *context-mediated behavior*, or CMB. We will look at its origins, as well as its evolution when applied to different domains over time, with an eye toward lessons learned not only for CMB, but for contextual reasoning in general. We will then look at current and future work on contextual reasoning in real-world systems.

5.2. Context-mediated behavior

Context-mediated behavior arose as a reaction to the prevailing approaches to case-based reasoning in the late 1980s. In those approaches (e.g. Kolodner et al. 1985), a record of a problem-solving session – a case – was stored for later use.

When a similar problem-solving episode was later encountered, the reasoner would be "reminded" of the previous case, which it would then use as a source of knowledge about how to solve the current problem, based on how it was solved previously.

One strength of CBR is providing shortcuts to current reasoning by reusing work done in the past. Another strength is that it is useful in situations where there is little domain knowledge from which to derive *de novo* solutions. By recording what worked (perhaps serendipitously) and what did not work in the past, a case-based reasoner can solve problems even when it would have difficulty creating a new solution.

CBR is thus useful in domains, such as medicine, law and other complex real-world domains, where although there is some knowledge about causality, actions to take, etc., much human reasoning is based on what worked or happened in the past. This is the case for domains where there may be a great deal of formalized procedures, but where workers "on the ground" may apply those procedures differently based on the situation using tacit knowledge rooted in past experience. CBR can therefore be seen as an approach with some of the same goals as Brézillon's (e.g. Brézillon and Pomerol 1999) work on *contextual graphs* (CxG).

A problem with CBR, however, is that cases do not really represent contexts – or rather, they conflate contextual knowledge with problem-solving knowledge that may be context-independent, as well as with knowledge about the situation in particular that is not useful to remember about the context in general. For example, a medical reasoner diagnosing a patient with recurrent pulmonary embolism may know a great deal about the situation once it has solved the problem; some knowledge would have come from the general knowledge used in the episode (pulmonary embolism is dangerous, we can expect the patient to present chest pain, shortness of breath and a cough, etc.), from the particular patient (they have a cough, they are short of breath, they have a history of deep vein thrombosis, their grandmother suffered from gout, their mother had breast cancer, etc.), from portions of the context that do not involve the patient (the reasoner is located in a hospital, the time of day is afternoon, the computer on which the reasoner is running is busy today, etc.), and from problem-solving in general (hypothetico-deductive reasoning is a good way to diagnose diseases, the reasoner can use a targeted questioning approach since it is an expert at pulmonary diseases, etc.). Not all of this information is actually useful in future situations like this that occur in the same kind of context – a patient's mother suffering from breast cancer has little to do with recurrent pulmonary embolism, for example. Since context is essentially compiled in the case along with the reasoning, the contextual knowledge *could* perhaps be reconstructed when the case is recalled for reuse, but this may involve an arbitrary amount of reasoning effort that would have to be repeated *each time* the case is used. Worse, some of information in the case can actually lead the reasoner astray: for a new patient suspected of having breast cancer, the fact that her mother had it may cause the reasoner to remember this case rather than a more salient one that would lead to a correct diagnosis.

These problems arise from there being no real reasoning about context nor attention to identifying contextual knowledge that can help solve future problems; processing occurs as cases are recalled, interpreting them in light of their similarities and differences from the new situation. While this is a strength of CBR – a case can potentially be used when we simply do not know what *is* salient and what is not – it is also a significant weakness.

The MEDIC project (Turner 1989, 1994) aimed to address these problems. It was observed that, while cases were indeed important in its domain of medicine, they seemed to not be used as much as more general knowledge. For example, upon seeing an unusual case, while a doctor *might* remember another past case with the same contextual features (patient's signs and symptoms, for example), this did not seem to happen as often as the doctor recalling a *generalized case*, something along the lines of "usually when I see patients with these symptoms, I think of histoplasmosis". While this could be explained as the doctor being reminded of multiple cases, from which they generalized on the fly, a more parsimonious explanation is that they instead drew from general knowledge built up from multiple cases.

The kind of memory often used for case-based reasoners, a so-called *dynamic memory* (Schank 1982) of the form implemented by Kolodner (1984), happens to create something that is very much like generalized cases as a by-product of storing cases. A new case is stored by traversing indexing structures that organize the memory, which are themselves associated with knowledge structures called *memory organization packets* (MOPs). Indices are chosen by selecting similarities and differences between the new case and MOP, which acts as a kind of synopsis of the cases and other MOPs it indexes, and it is built and updated by inductive reasoning as cases are added. The chief insight in MEDIC was that it is often better to reason from these generalized cases when one exists that matches the new problem, rather than from individual cases.

Not all of the knowledge in these MOPs was of the same type, however. Some concerned problem-solving procedures (inference rules, goal decompositions, plans, scripts) that were not unique to the MOP, but were rather often instantiated in the situations it represented. The remaining knowledge was more about the context in which that knowledge was useful, as well as predictions about features of the world that would exist in the context represented by the MOP and in the situations it indexed, which were really instances of the context having occurred.

A new style of reasoning, called *schema-based reasoning* (SBR), arose from those observations (Turner 1994). Reasoning in SBR is guided at all times by two kinds of schemas: *procedural schemas* (p-schemas), which represent scripts, plans or goal decompositions that may be useful in many different contexts; and *contextual schemas* (c-schemas)[2] that represent contexts in which the reasoner may find itself.

SBR separates procedural knowledge from contextual knowledge, allowing them to be reasoned about and applied separately. For MEDIC, contextual knowledge, in the form of c-schemas, provided predictions about features that are expected in the context, including symptoms the patient may have, salient patient characteristics and so on. C-schemas also provided information about which p-schemas would likely be useful for achieving goals in the context.

MEDIC provided explicit ways that contexts are related to one another as determined by its memory. The c-schemas were memory organization packets, so they had the dual role of representing contexts as well as organizing the memory. C-schemas were organized, as in Kolodner's (1984) work, in a tangled generalization–specialization hierarchy, essentially a number of overlapping discrimination networks automatically created as new cases were added. By traversing these using features of the current situation, the best (i.e. most specific) c-schema, and thus, contextual knowledge, was used for the situation.

Procedural schemas fulfill some functions similar to Brézillon's (2003) *proceduralized context*, that is, contextual knowledge that has been instantiated in a particular way according to the situation and goals (although no such connection was made at the time). P-schemas can be viewed as pieces of c-schemas, similar to the *scenes* of MOPs (Schank 1982), and they are stored in their own generalization–specialization hierarchies in memory, just like c-schemas. A specialization of a p-schema is just a version of that p-schema as it has been or should be used in a particular context.

Work on MEDIC showed that explicit context representation was useful for capturing different kinds of diagnostic contexts and guiding the reasoner to apply the appropriate knowledge at the right time. It generalized CBR to make use of general

2 Originally called *diagnostic schemas*, but later renamed when it was realized what they really represented.

kinds of problem-solving "cases", while still preserving true case-based reasoning if no such stereotypical generalized knowledge existed[3].

While work in the MEDIC project provided a starting point for a model of how to represent and use contextual knowledge in a complex domain, it was far from a general model of context-sensitive behavior. For one thing, MEDIC was a very simple version of what would later be termed a context-based intelligent assistant system (CIAS; see, for example, Brézillon (2014)). It did not deal with the full range of knowledge and flexibility needed by, for example, an autonomous agent.

5.2.1. *CMB for autonomous agents: Orca Project*

The Orca Project, which began in 1990, extended the work on schema-based reasoning begun in MEDIC to address the needs of an intelligent autonomous agent performing complex tasks in real-world environments: an autonomous underwater vehicle (AUV). It is from this project that context-mediated behavior (CMB) arose.

Ultimately, work on Orca led to a more capable architecture for a context-based intelligent agent, as shown in Figure 5.1. Orca has a schema applier and agenda manager that were present in MEDIC and responsible for expanding p-schemas and maintaining the reasoner's focus of attention, respectively. In addition, there is now an event handler whose job it is to examine incoming percepts and update the reasoner's working memory, while monitoring for both expected and unexpected events in the world. A long-term memory holds both c-schemas and p-schemas, and it is now a semi-autonomous unit that continuously monitors working memory to find potentially relevant c-schemas and p-schemas, which are related to the reasoner as "remindings" (see Kolodner (1984)).

Finally, there is now a dedicated context manager, ConMan, which is responsible for assessing the current situation to determine what the context is, as well as collecting the relevant contextual knowledge and seeing that it gets to the other modules that need it. Orca interfaces with the rest of the agent's software and hardware via its low-level architecture interface that can be changed depending on what Orca is controlling. The behavior of all of the modules, with the except of long-term memory, is context-dependent and relies on ConMan for knowledge about what is appropriate in the current context.

3 Interestingly, CBR was implemented as procedural schemas in MEDIC, as was partial-order planning (POP) in a later reasoner (Ramakrishnan and Turner 1997). Thus, the reasoner could choose the overall kind of from-scratch or case-based reasoning to use when no context-specific p-schemas were known, or a c-schema could directly suggest a particular kind if that was best in the context.

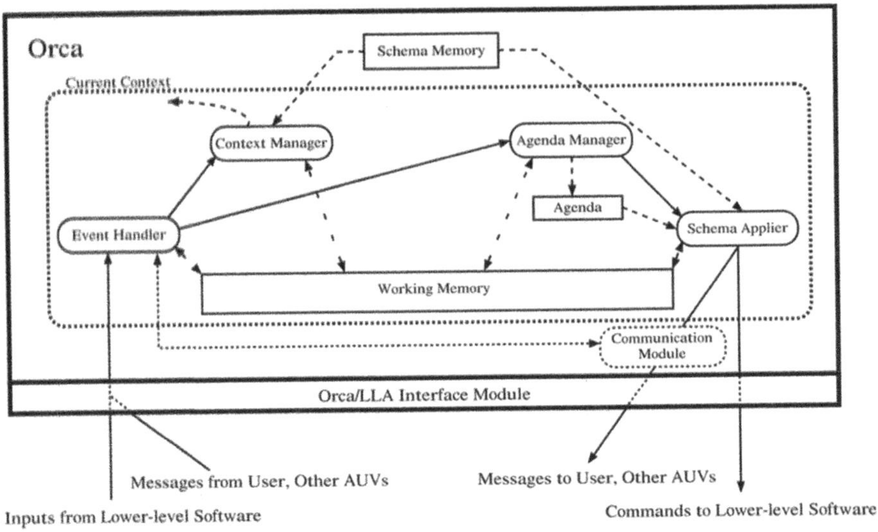

Figure 5.1. *Architecture of the Orca program*

5.2.2. *Contextual schemas*

We also firmed up our notion of what a "context" is with Orca. To start, we define the agent's *situation* as comprising all the features of the world, the agent and the problem; as such, a situation is potentially infinite in extent and content (see McCarthy's (1993) notion of context), and, of course, an agent only knows and represents a portion of its situation. Each situation in a complex domain is likely unique; as Heraclitus noted, no one can step in the same river twice, since something (the river, the person) will have changed between the two steps.

However, that does not mean that there are no functionally equivalent *classes* of situations with respect to the agent. For example, the situation when I brushed my teeth this morning differed from the corresponding situation yesterday: there were different pets in the bathroom, there was almost certainly a different amount of toothpaste on the brush, the day of the week was different, the phase of the moon was slightly different, Mars was in a different place in its orbit and so on *ad infinitum*. That does not, however, change the fact that the two situations share most salient features, and most inferences, heuristic and procedural knowledge about one still pertains to the other.

This is in general what we mean by a *context*: a class of situations that, in some important senses for the agent, are equivalent in their predictive and interpretive power. Any given situation, then, can be classified as an *instance* of some context. Both of the situations above are instances of me having been in the "tooth-brushing" context, for example.

In fact, almost all situations can be viewed as instances of more than one context. For example, consider an autonomous underwater vehicle that is searching for a shipwreck under arctic sea ice, when it notices that its power has dropped below a safe threshold for further operation. Although we could – and would – consider this context to be something like "searching for a shipwreck under sea ice with low power", this may be the only instance of such a specific context that the agent has ever encountered, and so there may not be much that can be learned trying to recognize the entire context. However, we can also consider this situation to be an instance of (at least) three contexts: searching for a shipwreck, operating under sea ice and experiencing low power. It is much more likely that there *is* prior knowledge about one or more of these contexts that can inform the agent how to behave in its current, perhaps novel, situation.

Thus, from the pragmatic point of view of which contexts make sense for a reasoner to represent and reason about, we consider the answer to be when a class of situations (a context), all similar enough to have the same (or nearly the same) predictive and interpretive power – and, thus, the same implications for an agent's behavior – cannot be readily synthesized from other known contexts. Thus, unless there were some novel implications for behavior in the AUV's context above that cannot be captured in the merger of knowledge about the three component contexts, we would not represent it, but rather synthesize it from the components as needed.

If there *were* something unique about the combination, however, we would see that as reason to remember the particular context. For example, we can imagine a person having a good amount of knowledge about the context of eating a meal, and a similar amount about the context of driving. Here in the US, we often do both at once (probably to the detriment of each). However, combining our knowledge of the two contexts here will *not* give us the full picture, as unlike the usual eating context, we have no table and food is unlikely to remain where we put it in any case, and unlike the usual driving context, we will not have both hands on the wheel. Thus, in this instance, we would expect the reasoner to have knowledge about this particular context, including that "eating while driving" is indeed a distinct context. It would be related to, but distinct from, the other contexts.

Each context in our approach is represented as a contextual schema. Unlike in MEDIC, there will likely be more than one c-schema used at a time to represent the entirety of the agent's current context, as we just discussed. The question, then, is what should these c-schemas contain? That is, what kind of knowledge is important for the agent to have about contexts in which it may find itself?

The answer is "it depends" – it depends not so much on the domain as on the reasoner itself. Here, we will discuss the general kinds of contextual knowledge we have found useful for an intelligent autonomous agent, such as an AUV mission controller, to have. In later sections, we will discuss how the kinds of knowledge differ for agents using different reasoning styles or with different capabilities (e.g. collaborative agents, deep-learning agents).

In CMB, as it is implemented in Orca, contextual schemas contain several new (with respect to MEDIC), different kinds of knowledge that can be broadly classified into prescriptive and descriptive knowledge.

Descriptive contextual knowledge. *Descriptive* contextual knowledge is just that: it describes the context. This includes knowledge about features that have to be present for the situation to be considered an instance of the context represented by the c-schema, and also features that the agent can expect to be present at various points while in the context. For example, for a c-schema representing "searching for a shipwreck" to be considered as an AUV's current context, the current goal would need to be to search for an object and for that object to be a shipwreck, the AUV would need to be in the water (as opposed to on its support ship's deck) and so forth. Features that should be predicted by the c-schema would include that the AUV is moving, power is decreasing, objects will be detected, one of the objects detected may be a shipwreck and so on.

Another type of descriptive knowledge that we have found useful is the context-dependent meaning of concepts. Orca is a symbolic reasoning agent (as opposed to a neural net-based agent), and so various symbols or predicates are used to represent concepts. For example, `(depth $self nominal)` might represent the belief that the AUVs depth is neither too deep nor too shallow. However, the meaning of "nominal" is different in different contexts: in a harbor, it is important to avoid surface vessels and debris on the bottom, and so "nominal" may be anywhere in a depth envelope of, say, 3 m to 10 m (depending on the harbor, of course); when operating in the open ocean, "nominal" may be anything from a shallow depth that avoids the current wave action to the vehicle's maximum safe operating depth (the "crush depth"), and when operating in the open ocean when there is a hurricane overhead, the top of the "nominal" depth envelope should be deeper.

We could, of course, just avoid the context-dependent meaning of "nominal" by using actual depth values. However, there is good reason to keep the symbolic representation, even though its actual meaning changes: there are other aspects of "nominal depth" that have implications for the agent's behavior. For example, if the agent realizes that it is no longer at a nominal depth, it should do something to restore that state (go down if too shallow, go up if too deep), and this is the same no matter what "nominal" means in terms of actual depth values.

We handle this in CMB by storing context-dependent meanings in c-schemas when they are important for how the agent should behave in the context. Keeping with our example of "depth", in one version of Orca, depths were represented as fuzzy values (Zadeh 1979), where each value is represented as a membership function, mapping numeric depth values to membership in the corresponding set. For example, Figure 5.2(a) shows one set of membership functions for the fuzzy sets "too shallow", "shallow" and so forth, for some harbor. A depth of 20 meters would thus be considered to be "too deep". Part (b) of the figure shows another set of membership functions, this time for the open ocean. Here, a depth of 45 m would have the same meaning as 20 m did in (a) and should trigger the same action, even though the actual numeric depth values are quite different. CMB handles this by storing the membership functions defining concepts such as these in the c-schemas where they are relevant.

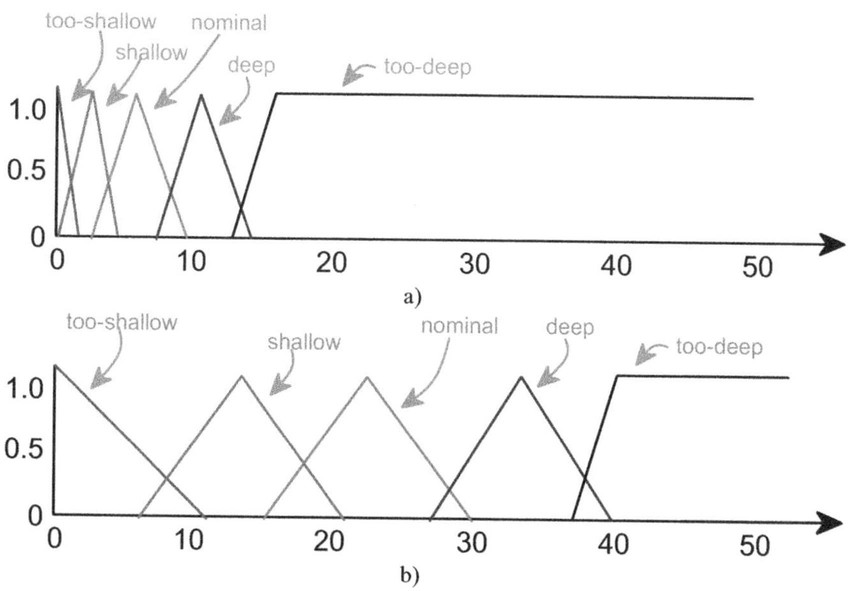

Figure 5.2. *Example fuzzy depth values for two different contexts. For a color version of this figure, see www.iste.co.uk/brezillon/context.zip*

Descriptive contextual knowledge is used by all modules in Orca, but in particular by the event handler. Context-dependent meanings and predictions are important for this module to be able to update the working memory, as well as identify events that have happened.

Prescriptive contextual knowledge. The other kind of knowledge stored in a c-schema is *prescriptive* knowledge, that is, knowledge about how to behave in the context. The exact kind of prescriptive knowledge stored will depend on the agent's reasoning techniques, but we can identify some general kinds.

First and most obvious is goal-associated prescriptive knowledge, that is, knowledge about context-appropriate ways to achieve goals. For Orca, which is a schema-based reasoner, this is knowledge linking goals to procedural schemas that are useful for achieving them in the context[4]. For other reasoners, as we see below, these may be rules, neural network topologies and weights and so on. This kind of procedural, prescriptive knowledge in Orca is used by the schema applier as it looks for and expands p-schemas to achieve mission goals.

A second kind of prescriptive knowledge is knowledge about how goals are related to one another in terms of importance and urgency in the context. This attention focusing knowledge helps Orca determine what to work on during a context. For example, it is reasonable to expect the goal "surface" to arise when an AUV controller realizes it is about to lose power. In most contexts, this is a reasonable thing to do. However, if the AUV is on a rescue mission and has just located a diver, it should likely not do this, but rather continue with the mission in the hope that it can help the diver before its power is completely depleted. This kind of contextual knowledge is used by Orca's agenda manager to determine which goal to work on at a given time and which to delay or ignore.

A third kind of prescriptive knowledge concerns event handling. How an event presents in a given situation depends on the context. For example, in the open ocean, a sonar contact could be many different things, including large fish, a whale, another submersible and so forth, and there may not be much else that can be inferred other than "object at (x,y,z)". However, in the context of being in a mine field, a sonar contact should be considered evidence for the object being a mine, given not only the top-down contextual knowledge, but also the importance of misidentifying a mine (better a false positive than a false negative). Other context-specific knowledge might include how to gather evidence to confirm the event, as well as what the event might mean for the agent in the context. For example, an object in front of an AUV

4 To be more precise, the link is from a goal to a p-schema that can then be specialized by the reasoner using the p-schema's memory organization indices to find even more specific p-schemas.

during docking means something quite different with respect to the effect on the vehicle than if the context were instead a mine field. Finally, what to do about the event is likely context-dependent; while the agent may be able to determine this based on the projected effect of the event, it may make sense for a c-schema to directly provide a reaction, for example, in the case where there is insufficient information to link projected effects to actions, there is a conventional response, or when response time is critical. Going back to an earlier example, if an AUV notices the event "power failure", in most contexts, it makes sense to abort the mission, surface and ask for help; indeed, this may not even need to be represented in most contexts, but rather it should be a default behavior. This – activating a goal to surface and radio for help – would even be a reasonable response for a rescue context, since in that context, attention-focusing knowledge should keep the goal from being worked on until a diver, for example, is found. When operating under sea ice, however, this response is *not* reasonable; instead, the agent's event handler should instead activate a goal to get out from under the sea ice first; in a harbor, where the vehicle could easily collide with surface vessels, the contextual knowledge should perhaps instead tell the event handler to activate a goal to land on the bottom and release a buoy so the vehicle can be found and retrieved.

A fourth kind of prescriptive knowledge has less to do with taking actions or changing goal priorities than modulating the overall behavior of the agent. We call this kind of knowledge "standing orders"[5], and it mostly consists of parameter settings appropriate for the context, although it can sometimes be useful to activate/deactivate goals as well. This kind of modulation is seen in human context-dependent behavior; for example, when entering a library, an individual lowers their own voice automatically, without conscious thought. Standing orders give an artificial agent this kind of automatic, context-specific behavior modulation, too. For example, they could allow an AUV entering a harbor to automatically tighten its depth envelope to avoid the surface and the bottom, or to turn off equipment that may trigger mines when entering a mine field.

By way of a summary, Figure 5.3 shows an example c-schema from our intelligent AUV mission controller domain. In section 5.4.2, we describe an ontology for contextual schemas and contexts themselves that arose from a related project in our lab.

5 Thanks to the late D. Richard Blidberg, of the Autonomous Undersea Systems Institute (AUSI), for the name (as well as for so much more – he was a mentor to the author – you are missed, Dick).

```
Actors:
  ?self = an instance of ^AUV

Objects:
  ?place = an instance of ^PLACE
  ?mission = an instance of ^MISSION
  ?wc = an instance of ^WATER-COLUMN with DEPTH = ^SHALLOW
  ?surface = an instance of ^WATER-SURFACE with TRAFFIC = ^HIGH-TRAFFIC
...

Description:
  ?context = this ^C-SCHEMA, name ?context = "in harbor"
...

Definitions:
  (^FUZZY-DEFINITION-DESC
    (LINGUISTIC-VARIABLE (SLOT ^PHYSICAL-OBJECT DEPTH))
    (LINGUISTIC-VALUE SHALLOW)
    (MEMBERSHIP-FUNCTION ((0 0) (2.5 1) (5 0)))
    (CF 0.8) (COMBINATION-TYPE REPLACE))
    ...

Standing orders:
  (^STANDING-ORDER (CONDITION T) (CF 0.8) (WHEN DURING)
    (DESCRIPTION (SET-LLA-PARAMETER DEPTH-ENVELOPE (5 10)))
...

Events:
  (^EVENT-DESC
    (DESCRIPTION (POWER-LEVEL ?self LOW)) (DIAGNOSTIC-INFORMATION NIL)     (LIKELI-
HOOD UNLIKELY) (IMPORTANCE CRITICAL))
    (EFFECTS ((^EVENT-DESC (DESCRIPTION (STATUS ?mission FAILED) (CF 0.9))
              (^EVENT-DESC (DESCRIPTION (STATUS ?self FAILED) (CF 0.9))))
    (RESPONSE (^RESPONSE-DESC (DESCRIPTION (ACHIEVE (^A-ABORT))) (CF 1.0)))
    ...

Goals:
  (^GOAL-DESC (DESCRIPTION (^ACHIEVEMENT-GOAL (STATE (AT ?self (?x ?y 0))) (IMPORTANCE LOW))
...

Actions:
  (^ACTION-DESC (DESCRIPTION (^A-ABORT)) (ACTION (^P-ABORT-TO-BOTTOM)) (CF 0.9))
```

Figure 5.3. *Part of an example contextual schema representing being in a harbor. Some of the actual knowledge representation has been replaced to make it easier to understand*

5.2.3. *Context assessment*

In order to use the appropriate contextual knowledge for a situation, the context manager, ConMan, must perform *context assessment.* In early work on MEDIC, this was relatively straightforward: the memory was probed with features of the current situation to find the most specific matching c-schemas (then called diagnostic MOPs), based on simple metrics based on the program's knowledge representation; then the most specific one of these was used as the source of contextual knowledge. However, it was realized while working on Orca that this was too simplistic; this, coupled with the recognition that a single situation could be an instance of multiple contexts, led to a new way of doing context assessment.

In CMB, we view context assessment as a fundamentally diagnostic task, similar to the way a physician diagnoses a patient's current problem (i.e. the situation) as an instance of a known disease (which can be viewed as a class of similar instances of patients having the disease). "Symptoms" of the context correspond to features of the current situation, while the "disease" corresponds to a context as represented by a c-schema. Given this view, ConMan assesses the current situation using an abductive diagnostic mechanism derived from work in AI in medicine, in particular, the INTERNIST-I program (Miller et al. 1982).

Figure 5.4 shows this process. The features of the current situation, as reflected in the current state of the working memory, are used to traverse the long-term memory's indexing structures to retrieve a set of candidate c-schemas (this can be done both intentionally, by ConMan, or automatically, by the memory itself). In keeping with INTERNIST-I's terminology, we say that the features *evoke* this set of c-schemas. The c-schemas are then separated into *logical competitor sets* (LCSs; Feltovich et al. 1984). Each LCS contains c-schemas that roughly account for (explain the presence of) the same set of features in the current situation. The LCSs are scored based on the features they account for and how strongly they predict them. The highest-scoring LCS is chosen to be worked on first, and additional predictions from the members are used to *separate* them from one another in a process of differential diagnosis. This may include performing inferences to determine new facts derivable from those already known, or in principle (we do not yet do this) causing the reasoner to perform actions to gather more information about the world. During this process, the LCSs may need to be re-formed as additional information is gained. Ultimately, however, the topmost LCS will be "solved" when one context hypothesis (i.e. when one c-schema) outscores its peers by some threshold. At that point, the c-schema becomes one of the ones used to characterize the current situation, and ConMan moves on to the new topmost LCS.

The end result of all of this is a set of c-schemas, each of which accounts for some (possibly overlapping) set of the situation's known features. Each represents a diagnosis: the current situation is a member of the class of situations represented by the c-schema. All of the c-schemas are then merged into a coherent representation of the total context. We have referred to this using several different names over the years, but here we will just call it the *context object*; in essence, it is a new c-schema that may or may not be stored in long-term memory after leaving the context, depending on whether or not there is something new about it that cannot be re-created from the c-schemas it was created from.

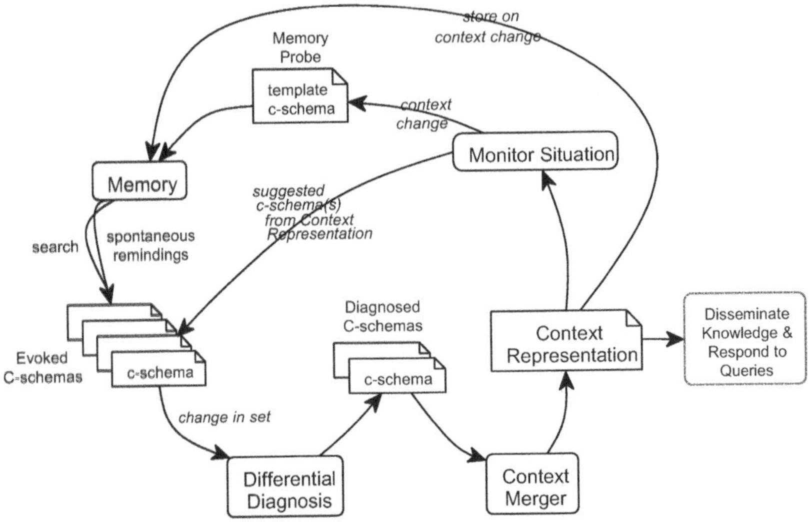

Figure 5.4. *Context assessment*

As the situation changes, one of the things ConMan watches for is to see if the situation has changed enough that one or more of the c-schemas comprising the context object are no longer are a good fit, or if better c-schemas have been found and returned by the long-term memory module. If so, then it decides that the context has changed, assesses the context again, and possibly stores the prior context object as a new context it has learned about.

5.3. CMB and planning

A problem with CMB as it has so far been implemented is that although it is meant to be implemented in a planning system such as Orca and provides planning and acting information to other modules, the CMB process itself is *reactive*: contexts are only identified once the agent has entered a situation that is an instance of the contexts. No provision is made for predicting future possible contexts, which could predict future situations and consequences, and thus guide an agent in its choice of actions. Thus, although a CMB-based AUV might be able to recognize when it enters a mine field unintentionally and thus behave appropriately while in it, it would be up to the planning component to avoid the minefield in the first place – even though the agent has entered the mine field previously in a similar situation and so *should* have been able to prepare before entering it. However, it would wait until it enters the context to implement changes such as turning off sensors that may trigger a mine explosion. There is thus a brief period of time, while context assessment occurs, when the AUV is likely to do the wrong, that is, context-inappropriate, thing[6].

Future contexts are also important to consider when deciding how to behave in the *current* context. When in the context of planning a vacation, a person can think about what they want to do when they are at their destination: where they would like to stay, what activities they might do, what kinds of food they may eat and so forth. This is important, not least of which so that they can decide what to pack, how much money to budget for the trip and determine whether they can afford it. Although this could be handled by some sort of fine-grained plan projection mechanism, it is more reasonable to view it as the person imagining future contexts they may be in, and then using general knowledge they have about those contexts to predict what they need to do now.

One early attempt to extend schema-based reasoning, including context-mediated behavior, to accomplish this sort of "context projection" was the ACRO project (Turner et al. 2005; Albert et al. 2007). This was based on earlier work (E.H. Turner, 1993) by one of the members of our lab on devising ways to structure a conversation using "conversation MOPs" (Kellermann et al. 1989) to predict future contexts in the conversation. This allowed a planner to identify where a future portion of a plan needs to be expanded to provide enough contextual information to make current decisions, including decisions about where to place new incoming goals in the evolving plan.

A more recent, ongoing project in the lab is aimed at directly integrating knowledge about possible future contexts into current behavior, and thus extending CMB. This approach is called *context-driven behavior* (CDBr; see Chapter 7 and Wilson and Turner (2015)) because contextual schemas are used as a primary knowledge source to drive the behavior of the system toward future contexts in which goals are satisfied rather than just providing knowledge to help behave in the current context.

Wilson describes several kinds of contexts important for CDBr in Chapter 7. *Environmental contexts* pertain to the environment an agent finds itself in. These are rather simple (e.g. "daytime") and "rigid" in the sense that there is nothing much the agent can do to modify the context itself. They constrain behavior in much the same way contexts do in CMB so far. *Agent-related contexts* are similar, but pertain to properties of the agent itself.

Problem-solving contexts (see Figure 5.5) are different in that they come into being as a result of goals and plans the agent has committed to, and they predict

---

6 Example from Wilson in Chapter 7 of this book.

resources that may be needed and future features of the situation that may occur due to the planning/acting process. As with some aspects of CMB's c-schemas, contexts such as these that are goal-related can help an agent distinguish between a context in which it is carrying out an action (e.g. in one of Wilson's examples, cooking a meal) and the one where the same features may be present (e.g. stove is on, food is out), but it is *not* the actor (and so the agent should not just take over cooking).

Problem-solving contexts are unique because they are used to prevent a premature exit from the context, or at least exiting without taking appropriate actions to cleanly leave the context. For example, an agent knowing that their problem-solving context is "cooking a meal" should prevent them from just leaving to satisfy another goal (e.g. going to pick up the children from school) without first turning off the stove, putting any food away that may be on the counter, etc.

```
Actors:
    self
    *others ;;zero or more intelligent agents

Setting:
    Day: Saturday
    Time: 12:30PM
    Location: (and (at-home self)
                   (at-kitchen self)
                   (at-counter self))
    Environment: (and (is-on stove)
                      (in-pan sandwich)
                      (on-stove pan))
    Agent: (is-hungry self)

Problematic features:
    Feature: (is-on stove)
      Clean-up action: (turn-off stove)
    Feature: (on-stove pan)
      Clean-up action: (remove pan stove)

Feature constraints:
    Feature: (at-home self)
        Constrained-by: (and (is-on stove) (on-stove pan))
        Preference: 1
    Feature: (at-kitchen self)
        Constrained-by: (and (is-on stove) (on-stove pan))
        Preference: 1
    Feature: (at-stove self)
        Constrained-by: (and (is-on stove) (on-stove pan))
        Preference: 0
```

Figure 5.5. *The problem-solving context "cooking on the stove" (courtesy Wilson)*

Problem-solving contexts can give rise to *persistent contexts*, which ensure that the agent keeps some feature of an over-arching context in mind until it can be taken care of, even after the current problem-solving, environmental, or agent-related contexts are no longer in effect. To again use one of Wilson's examples, an agent's problem-solving context while grocery shopping may generate a persistent context of having perishable foods in the car. Recognizing this context would keep the agent from, for example, stopping for lunch with friends on the way home, since that would lead to the food spoiling.

In CDBr, these kinds of contexts can all be reasoned about and brought into focus to help the agent plan for entering and leaving contexts appropriately, both the current one as well as contexts in the future.

5.4. CMB in multiagent systems

5.4.1. *Context-appropriate organization and reorganization*

The CoDA (Cooperative Distributed AOSN control) Project in our lab examined ways to control an open, heterogeneous multiagent system (MAS) comprised of autonomous and semi-autonomous intelligent systems operating over long time periods (Turner and Turner 2001, 2004). The target system was an AOSN, an autonomous oceanographic sampling network (Curtin et al. 1993) comprised of AUVs, moorings and other agents (see Figure 5.6). The system is *open* in the sense that its composition is expected to change over time as agents leave (due to hardware failures, needing to refuel or being needed elsewhere) or enter the system. The tasks undertaken by the system can also change over time. The problem then is how to organize the system effectively to carry out tasks in its current context as well as to reorganize when the context changes.

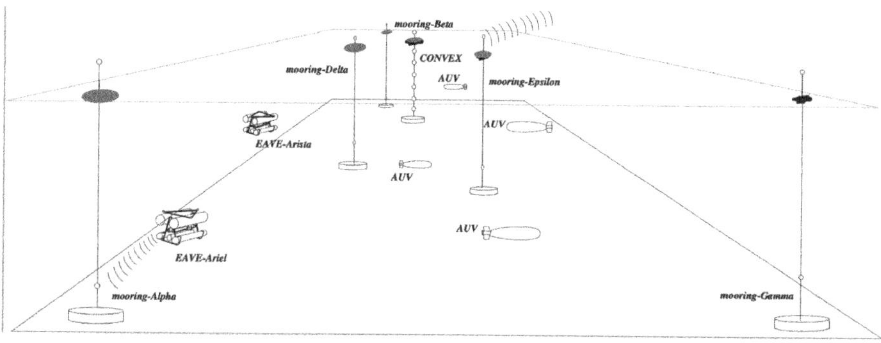

Figure 5.6. *Sketch of an AOSN. For a color version of this figure, see www.iste.co.uk/brezillon/context.zip*

Much of the work on CoDA on problem-solving protocols and so on is not relevant to the current discussion. However, work on the project did extend CMB into the realm of multiagent systems, not only for controlling individual agents, but also for helping organize and reorganize the MAS itself.

The basic idea for CoDA is that an agent's context includes which other agents are present, as well as the usual contextual features that concern problem-solving. In addition, the entire MAS can be thought of as being in a *joint context* that also impacts and interacts with agents' own contexts. Individual contextual knowledge is needed to allow the agent to solve its own tasks, as well as to communicate with other agents (e.g. by predicting a context-appropriate communication method and language). Knowledge of the group's context can be used by the agents to determine if variations in their tasks will impact the group and make predictions about future events and actions by other agents.

A major focus of the work was examining how contextual knowledge can be used to help select organizational structures for the MAS. Figure 5.7 shows the overall process of MAS-level control in CoDA. Some, though not usually all, of the agents are assumed to be able to take part in controlling the whole system. These self-organize at the start into a *meta-level organization* (MLO), and so these agents are termed MLO agents. Since it is likely that not only the task but also the composition of the MAS is unknown to begin with (e.g. some agents may not have arrived safely or may have been sent after the others that were already en route, etc.), the organization protocols governing MLO formation are simple and involve tasks such as discovering the total resources available, the tasks and so on; the MLO is designed to be a loose, cooperative organization that makes few assumptions about the capabilities present other than that some of the agents are intelligent enough to follow the protocols.

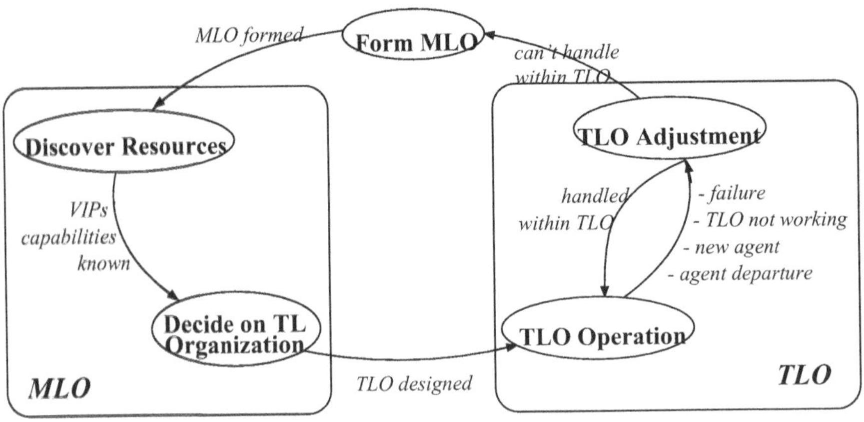

Figure 5.7. *Overview of CoDA's organization/reorganization mechanism (Turner and Turner 1998)*

The MLO itself, once formed, is tasked with designing an effective, efficient *tasklevel organization* (TLO) that will work on the task(s) until the context changes. Our approach to this is to use contextual knowledge, stored in agents' c-schemas, to suggest appropriate organizational structures, for example, a general hierarchy, a "product hierarchy", a committee structure or a contracting system (e.g. Smith 1980), as well as appropriate parameters for those organizations (e.g. number of layers in the hierarchy).

For this to occur, however, communication and negotiation is needed between the agents to determine what the joint context and relevant contextual knowledge is, since agents may not agree and/or only have some knowledge about various possible joint contexts.

In order for agents working together to communicate about contexts and contextual knowledge, they, of course, need a shared language; it does not matter very much what their internal representation is (first-order logic, frames, etc.), as long as each can translate to and from the shared language.

When there are multiple shared languages, determining which one to use is context-dependent, both for humans and artificial agents. For example, two native French speakers may choose to speak English if it is important that their friend, who only understands English, to understand what they are talking about. For artificial agents, properties of the environment or task may dictate which language they use. In the case of two AUVs operating underwater, they may choose a language such as COLA (E.H. Turner et al. 1994) that is optimized for low-bandwidth environments, whereas if they were at the surface, they may choose a more expressive language that requires fewer inferences and can represent additional concepts. Some organizations require different languages, too; for example, a market-based contracting system would require a language in which bids, awards and so on can be expressed. Although we have identified the need for context-dependent language selection, going further in this regard is a topic for future work. The expressiveness of the language will likely play a crucial role in interagent communication, including shaping that portion of each agent's model of the world derived from communication with others.

5.4.2. *An ontology for contextual knowledge and contexts*

In addition to a shared language, agents that are to communicate about their joint context need a shared ontology, both for contextual knowledge and for contexts themselves. Research in our lab focused on this problem in the CoDA project. This work not only resulted in an approach to multiagent context-mediated behavior, called MASCon, but also provided a knowledge representation and ontology that all of our projects can use to represent contexts and contextual knowledge (Turner et al. 2013; Rode and Turner 2015).

Prior work on CMB used ad hoc knowledge representations that were often idiosyncratic to each project. While this is sufficient for projects in which an agent works by itself, it is not sufficient when agents must communicate and work together, since understanding and generating whatever language is used requires each agent to have a representation whose syntax and semantics not only support the agent's problem-solving, but also support understanding the language.

We chose *description logic* (Baader 2003) as our representation language. It has well worked-out syntax and semantics and is widely used in AI, for example, in multiagent systems, NLP, and the semantic Web (Berners-Lee et al. 2001). In addition, there is good tool support, for example, the Protegé ontology editor (Noy et al. 2003) and the RACER theorem prover (Haarslev and Möller 2001).

The description logic on which we based our work is the L1 language (Teege 1994). This language provides operators for union and intersection of concepts, existential and minimal cardinality role restrictions, and role composition. It has the property of structural subsumption, which allows subsumption applied to concepts to reduce to subsumption tests of single clauses that cannot be further decomposed; this is important to MASCon's communication mechanism. The language also provides data types, much like the Web ontology language OWL-DL (McGuinness and Harmelen 2004).

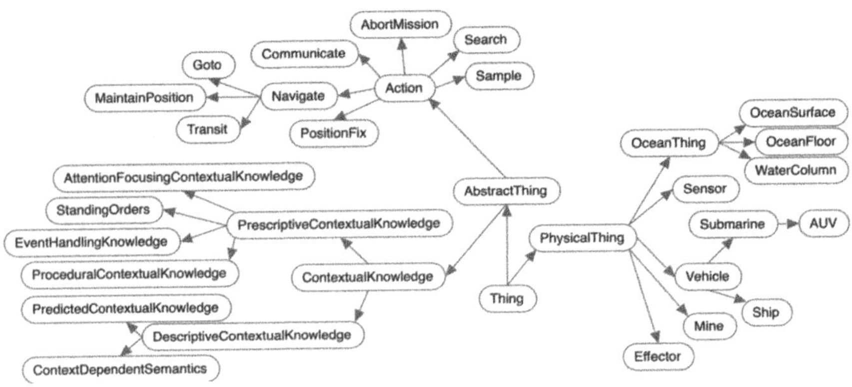

Figure 5.8. *A portion of MASCon's ontology (Rode and Turner 2015)*

All of the reasoner's knowledge, including all of its contextual knowledge, is defined using an ontology. Figure 5.8 shows a portion of the ontology (arranged to fit easily on the page); "Thing" is the top-level concept. To the left of the figure is the portion of the ontology that concerns contextual knowledge. Each of these concepts also has roles and definitions in terms of other concepts. For example, Event Handling Knowledge was defined to have roles:

```
(handlesEvent SOME Event)
(hasImportant SOME DefaultValuePartition)
(respondsWithAction SOME Action)
```

Contextual schemas contain assertions in the language. For example, a piece of event handling knowledge about power failure may be defined as the assertion:

```
(AND EventHandlingKnowledge
  (SOME handlesEvent PowerFailure)
  (SOME hasImportance High)
  (SOME respondsWithAction AbortMission))
```

```
BSM-1(0.60): (AND PredictiveContextualFeature
               (SOME expectsPresenceOf
                    (AND MissionArea (SOME hasFuzzyValue Broad))))
BSM-2(0.34): (AND PredictiveContextualFeature
               (SOME expectsPresenceOf (AND SamplingMission
                                         (SOME hasSamplingTarget Thing))))
BSM-3(0.41): (AND ContextDependentSemantics
               (SOME hasFuzzyFeature
                    (AND Broad
                      (SOME hasFuzzyMembershipFunction
                           (AND ShoulderFunction
                             (SOME hasLocalMinAt (AND Float 5.0 mi^2))
                                               (SOME hasLocalMaxAt
                                                    (AND Float 20.0 mi^2)))))))
BSM-4(0.59): (AND StandingOrder
               (SOME hasActivePeriod DuringContext)
               (SOME hasOperationalSetting
                    (AND ObstacleSensitivity (SOME hasAdvisedValue High))))
SM-5(0.064): (AND EventHandlingKnowledge
               (SOME handlesEvent SensorFailure)
               (SOME hasImportance High)
               (SOME respondsWithAction
                    (AND TransferData (SOME hasObject
                                        (AND Sensor
                                          (SOME hasStatus FailureImminent)))
                                          (SOME toObject MAS))))
BSM-7(0.55): (AND EventHandlingKnowledge
               (SOME handlesEvent SamplingComplete)
               (SOME hasImportance High)
               (SOME respondsWithAction Transit))
BSM-8(0.74): (AND ProceduralKnowledge
               (SOME definesAction
                    (AND Transit (SOME hasIndex Survey))))
```

Figure 5.9. *A portion of a c-schema representing a sampling mission. The format is label(CF): definition (Rode and Turner 2015)*

Metadata is allowed to be associated with assertions. This is used for things that are necessary, but difficult or impossible to represent in the language, and that are not needed for language-specific inferences (e.g. theorem proving). The primary example of this kind of metadata is certainty factor (CF) information. If CFs were part of the language, then two assertions that were identical, but differently believed, would not match each other.

Figure 5.9 shows an example of part of a c-schema as represented using this description logic-based representation.

In addition to an ontology for contextual knowledge, work on MASCon also explored an ontology for contexts themselves, so that agents could communicate about them. This is somewhat more difficult to deal with than an ontology for contextual knowledge, since an ontology for the latter does not change over time and can thus be common to all agents, while the set of contexts *does* change and is idiosyncratic to each agent, depending on their experiences. Consequently, a static taxonomy-like ontology is insufficient.

However, if we assume (as we do) that c-schemas are organized in a memory consisting of generalization/specialization hierarchies, then that memory itself can function as an ontology. While it is idiosyncratic to an agent and dynamic, we can assume that the agents can nonetheless be given a set of "top-level" contexts that will *not* change, at least at the time scale of problem-solving; other c-schemas, as they are learned, will be indexed from these in various ways. Agents can thus communicate about a given c-schema by referencing the nearest top-level c-schema(s) it is indexed from, along with a set of differences from it/them. This was used as the basis for a simple communication scheme that requires relatively little bandwidth when compared with sending *all* of a c-schema's contents, and yet which allows the communicating agents to have at least some references (the top-level c-schemas) in common as they reason about what their joint context is (see Rode and Turner (2015), for details).

5.4.3. *Trust in context*

An issue that arises in multiagent systems is *trust*. This is true even when the MAS consists of cooperative agents if one or more of the agents is self-interested as well as cooperative. This is especially true in an open multiagent system (OMAS), where agents may come and go over time and, hence, little may be known by the other agents about them, including how they are likely to behave over time.

In the CATS (Context-based Agent Trust System) project (Whitsel and Turner 2015), we looked at how the context of a multiagent system affects trust decisions and how a context-sensitive trust mechanism could be implemented for MASs.

To allow focusing on trust decisions, a domain was chosen that is simple, yet where trust is key: Liars Dice (Liar's Poker). In this game, each player rolls some dice to get a "hand", which is roughly like a hand in poker. Each player then takes a turn "bidding" about what its hand is: a pair, three of a kind and so on. The next player either accepts the bid or challenges it; if the bid is accurate, then the challenger loses a point; if the bid is a lie, then the bidder loses a point. Each player has to bid a hand that is higher than the previous, thus providing an incentive to lie.

CATS looked for abrogations of the social norms associated with the game, in particular for collusion among players. Properties of the situation were used to identify the current context; the c-schema representing it would then provide information to help the agent determine whom to trust and how much.

Since this was an initial foray into trust assessment, and one that would support empirical evaluation, the implementation lacked much of the complexity usually associated with CMB. C-schemas were not stored in a dynamic memory, for example, and the contents were CLIPS (Giarratano 1993) rules.

Experiments showed that context-aware agents perform better than other agents when the task environment involves context changes, and that by assessing the context as it changes, agents can change their trust strategies appropriately.

5.5. (Deep) learning in context

Most of the work on CMB – indeed, on context in general in AI – has been exploring how context impacts and can be used to affect the behavior of symbolic reasoners, such as automated planners. However, context also plays an important role in human and animal learning (see, for example, Balsam and Tomie (1985)).

There has, of course, been prior work on learning in context in AI; the case could be made for case-based reasoning being a kind of context-based learning, for example. However, there has been relatively little work using context to help a modern machine learning system – for example, a deep neural network – learn more effectively[7]. An exception is the work by Stein and Gonzalez (2014), which improved learning in a neuro-evolutionary system by using context-based reasoning (CxBR) to partition the learning task by context.

Consider a neural network that is tasked with learning when an autonomous underwater vehicle is too deep. As we discussed above, this is context-dependent: if the AUV is in the open ocean and on no particular mission, it is only too deep if it is near the bottom of its operational envelope. If it is in a harbor, then "too deep" is likely to be determined by the harbor's depth, rather than the AUV's crush depth, which is likely to be much deeper. If it is in either location and attempting to get a GPS fix, then "too deep" would be quite shallow. A neural network attempting to learn "too deep" across all of these contexts would have to learn the context-dependent meanings for all contexts. It would need a wide range of inputs overall, only some of which would be useful for all contexts. It would thus have to learn which inputs to pay attention to and which to ignore, based on examples on which it is trained. Doing this would likely require a substantial number of layers and/or nodes/layer, and it would require a large number of training examples to learn which inputs are important and how they relate to the evaluation of "too deep" in different contexts. In essence, the network would be learning about the contexts at the same time it learns how to assess depth.

A better approach is to train networks *in context* and to let something else – for example, CMB – worry about what the context actually is. In early work, we stored the salient network parameters for simple fully connected networks – enough to instantiate the network – in c-schemas: number of layers, number nodes in each layer, weights and so on. When given an example situation in which to learn, the context manager retrieves the appropriate c-schema[8], then it instantiates the network specified by the c-schema to learn from the example. As long as the context remains the same, the network continues to process examples. When it changes, however, the current network's weights replace the ones stored previously in the c-schema (if learning is enabled), and the process begins again with c-schema representing the new situation's context.

---

7 In its simplest form, a neural network consists of interconnected artificial neurons, each of which takes weighted inputs from others and produces an output based on the summation as modified by an activation function (e.g. the sigmoid function). The neurons are arranged in layers, with weighted connections only between neurons in different layers. Learning occurs when, roughly, an error function is computed based on the difference between the network's output and the desired output, then this error signal is back-propagated through the network, adjusting the weights. A "deep" neural network (or deep learning system) is one that has an input layer, an output layer and more than one ("hidden") layer between them.

8 Only a single c-schema was used to represent the context of a situation in that work.

We hypothesized that context-specific networks can be smaller, with fewer inputs: since a number of features that impact the assessment are *contextual* features, they do not change as long as the context itself does not change. Hence, the network does not have to take them into account and need not have inputs for them. Context-specific nets should be quicker to train to reach the same degree of accuracy, since they are smaller. We also hypothesized that context-specific networks can be more accurate than ones that are not. A problem with neural networks – or inductive learning in general – is over-generalization: the network may be good in a wide range of contexts, but not very good in any particular one. This problem can be ameliorated by increasing the number of layers or nodes/layer, but this increases both network size (obviously) as well as training time.

Our experiments showed that the smaller, context-specific networks do indeed generate on average fewer errors than larger, context-independent networks ($p \leq 0.05$) (Arritt and Turner 2003). We also showed that as the number of contexts involved increased, training time increased much more rapidly for the context-independent nets than for the context-dependent networks (see Figure 5.10). The size of the context-independent network (number of nodes) was also estimated to grow exponentially with the number of contexts. Although there will be more of the smaller nets as the number of contexts increases, that number will only grow linearly.

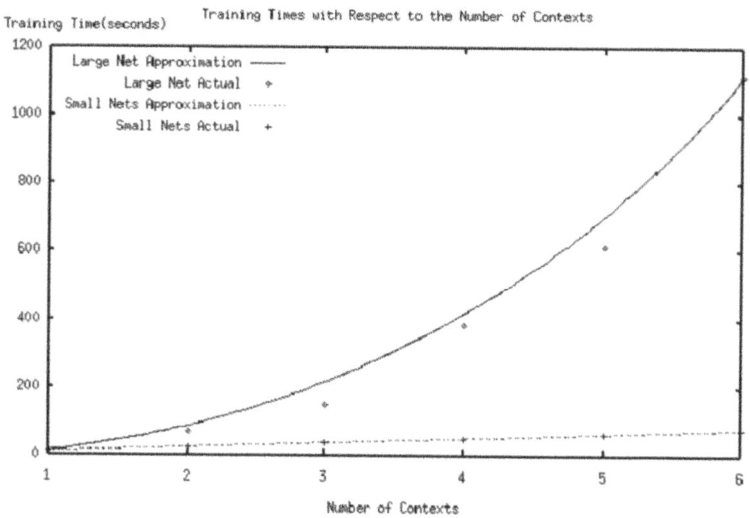

Figure 5.10. *Training time for large (context-independent) versus small (context-dependent) networks (Arritt and Turner 2003)*

Unfortunately, this early work only used fully connected networks with a single hidden layer – so not "deep learning" networks at all. We are extending this approach, now called Context-Dependent Deep Learning (CDDL), to complex, deep convolutional networks. This is motivated by a real need: to find and identify seabirds (gulls, terns, cormorants and eiders) and their behavior (nesting, roosting, flying) in aerial imagery.

Periodic aerial surveys of the islands off the coast of Maine are done by biologists to assess the number of nesting birds. Since seabirds are sensitive to the dynamics of their forage-fish prey, they are good proxies to assess the health of the coastal marine environment (Furness and Camphuysen 1997). Counting and identifying the birds is a very time-consuming task. Not only are there many birds per island, it is often difficult to find them (see Figure 5.11(a)). Processing imagery for a single island required approximately 14 hours each for two people to identify the approximately 2500 birds present. Given the large number of islands of interest (over 3000 in Maine alone), it is unreasonable to expect to have the human resources to count the birds on a regular basis.

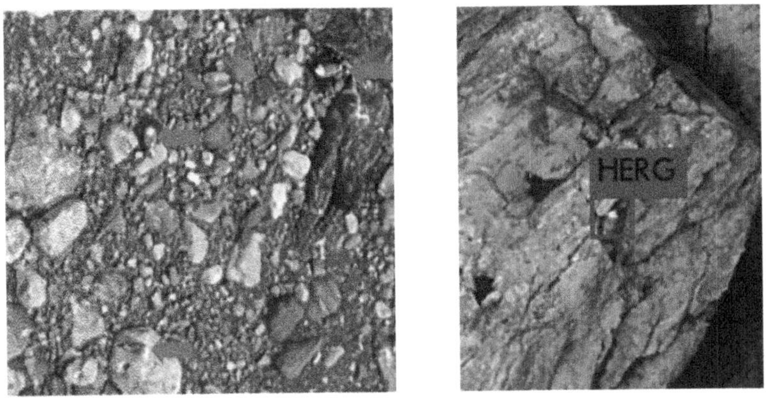

Figure 5.11. *(a) Portion of an aerial image showing locations of Herring Gulls (*Larus argentatus*). (b) A Herring Gull identified by YOLO. For a color version of this figure, see www.iste.co.uk/brezillon/context.zip*

Fortunately, recent years have seen a dramatic increase in the capability of AI-based image processing. Deep convolutional neural networks (CNNs; LeCun et al. 1989) underpin much of this. They are neural networks that are patterned loosely after mammalian visual systems. The network performs a series of functions (convolutions) on the input image to detect increasingly abstract features present in the image. This culminates with the network (or, usually, a fully-connected neural network last stage) being able to classify entire objects.

Detecting seabirds in images requires image segmentation and localization along with classification. *Segmentation* involves finding multiple objects in an image, and *localization* provides locations and bounding boxes for them. Many deep CNNs have been developed to do this. We chose a recent, very capable fully convolutional network, YOLO (You Only Look Once; Redmon et al. 2016) as the network upon which to base our work.

Figure 5.11(b) shows an example of output from YOLO in our domain. The network achieved accuracy on par with other image segmentation/classification networks for similar tasks. Unfortunately training times were very long, even on a state-of-the-art GPU cluster capable of several petaFLOPS (quadrillions of floating point operations per second). For example, one of the training sessions resulted in only a 60% recognition success after a *week* of training. What is also problematic is the large number of labeled training images required.

More complex tasks require more training and a larger network (number of layers and nodes/layer, and thus more weights to learn). Our pilot study, discussed above, suggested that for simple networks, eliminating the need for the networks to recognize the context decreases networks size in any particular context, while preserving performance quality. It also suggested that allowing networks with different structure, weights or both to be trained separately in different contexts can actually increase the network's accuracy.

Using a similar context-based approach for more complex networks in our current domain promises several benefits. First, it will reduce the network complexity and training time needed in any particular context; a network for detecting gulls among rocks and one for detecting them in grass, for example, are each likely to be simpler and easier to train than a single network that has to learn to recognize them in both contexts.

Second, it will allow effective allocation of training time and processing. For example, objects can be more difficult to detect in some contexts than others. A nesting gull is *much* easier to detect in grass than among rocks, and so it makes sense to use a deeper network or more training in the rocky context. With a single network that does not explicitly recognize the context, this is not possible. Similarly, if objects are known not to occur in some contexts, then there is no need to look for them there. For example, there is no need to look for nesting herons in rocky or grassy areas, nor to look for nesting gulls in trees. This will free up training time to look for the birds that *are* expected.

Third, by recognizing the context, training examples can be used more effectively. For example, one species of cormorant (the Great Cormorant, *Phalacrocorax carbo*) has been found on only a few islands in Maine, whereas the

Double-crested Cormorant (*P. auritus*) is much more widespread. A network trained on images of cormorants drawn from across all islands may not learn to recognize the distinction between the two, whereas if the known occurrence of *P. carbo* is taken into account to separate the examples, the networks trained for those islands are more likely to be able to detect both species.

In our domain, as in many others, contextual knowledge can be readily acquired from human experts. Biologists know a great deal about where in general they are likely to find or not find nests of different bird species, as well as the relative difficulty of identifying birds in different kinds of terrain. While this knowledge cannot easily be given to a neural network – human knowledge is symbolic, while a neural network's is sub-symbolic – it can easily be used by a symbolic AI program tasked with identifying the context.

Some features that we have identified as useful for characterizing contexts in this domain include the terrain (rocky, grassy, etc.), location (latitude/longitude, which island, etc.), altitude of the camera, weather (sunny, cloudy), time of day/year, type of camera, and whether the island is inhabited. These will be the features associated with contextual schemas and used to diagnose the context. The c-schemas for this problem will contain knowledge about the properties of networks that are appropriate. In some contexts, entire prior neural networks, including weights, should be provided; in others, it may be appropriate to create a new neural network, with its topology and kind of layers based on the system's past experience and knowledge from experts. Hyperparameters appropriate for the context can also be provided, for example learning rate and other properties of training, such as the number of epochs expected to be needed (related to the task's expected difficulty in the context).

Experiments are planned for the near future to determine if a context-based approach is useful in our domain and for complex neural networks. The control for these will be the current version of YOLO now being trained without regard to image context. It will be compared to smaller networks trained on image subsets drawn from similar contexts (e.g. grassy area vs. rocky shore, weather condition, etc.). The training time for the control network will be compared to the total training time for the smaller networks, and performance on test images will be compared.

Data will also be gathered about the overhead of context-switching. A complete CDDL system will almost certainly be presented images without regard for their context, especially as new images are acquired over time. It will have to assess the context of each image and, when the context changes, store current network properties and instantiate a new network (or at least load different weights). We will measure the time and space requirements when the context changes due to images being presented in a random order, which will also provide data about memory/disk requirements for CDDL in this domain. Other experiments will compare the performance of the networks under different frequencies of errors in context diagnosis.

After assuming the experiments show promise for using context-dependent networks, we will then proceed to implement CDDL along the lines shown in Figure 5.12. As can be seen, this is essentially just CMB with the ability to instantiate neural networks.

Issues that will need to be addressed include how to establish what is a context in this domain, how to identify and assess the contexts, and how to represent and store the weights and network topologies. In earlier work, this latter problem was simple, since there were not very many weights involved. However, YOLO can have many thousands of weights; in fact, the weights for one pre-trained version of YOLO v3 that we use occupy ~250 MB on disk. Not only is this potentially extremely space-intensive if there are many contexts, but since some or all of the c-schemas' weights will almost certainly need to be stored on disk, instantiating the networks is likely to be slow. We will investigate using ways of compressing the weights developed by a colleague in our school (Ganesh et al. 2020).

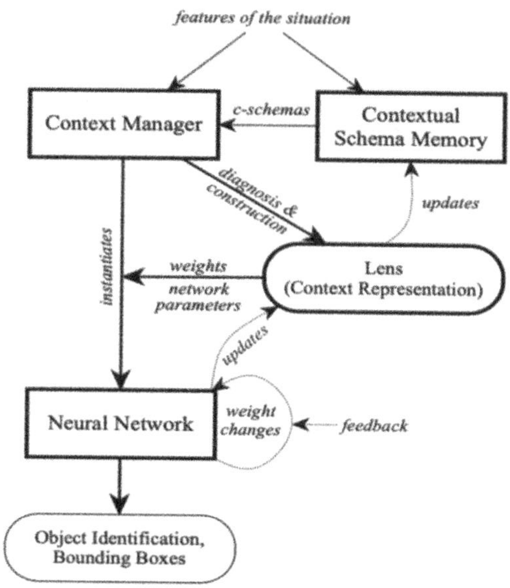

Figure 5.12. *CDDL process*

We anticipate that CDDL will be useful for a wide range of deep-learning techniques and applications.

5.6. Conclusion

In this chapter, we have described an approach to pragmatic, context-based reasoning, context-mediated behavior. Work on CMB has been done for several purposes and in different domains over the years, both demonstrating its widespread applicability, as well as developing and refining the approach. CMB has been applied in the domains of: medical diagnosis (in a rudimentary form), intelligent mission control of physical systems, multiagent systems and, most recently, context-dependent deep learning. Along the way, we have investigated context-specific event handling, attention focusing, procedural knowledge selection, behavior modulation, concept semantics, organization selection for MASs, and selection of neural network weights and topology. In addition, some initial work has been done on using contextual knowledge to drive behavior and predict future contexts, and on devising an ontology for contextual knowledge and, to some extent, for contexts themselves.

Very little of this work was formal and top-down, where a formalism is first devised, then its range of applicability determined. Instead, our work proceeded directly from the "scruffy"[9], pragmatic, side of artificial intelligence, where a solution for a particular, well-defined problem is devised, then that solution is examined to determine what general principles can be learned. As time progressed, however, we found, as expected, that it is better to use formal methods where they are appropriate *in service of the pragmatic needs of the work*. Thus, over time, we have incorporated into CMB fuzzy logic/sets, organizational design principles, deep neural networks, and description logic and its associated theorem proving.

We believe that this is the way forward for context-based reasoning in particular and AI in general: real problems driving solutions that may not fit any particular preconceived formalism, yet use formal methods as appropriate and needed. This, to us, is the *pragmatic* approach to context-based reasoning.

5.7. Acknowledgments

The work reported here spans 30+ years, and so it is not possible to thank all those who have contributed along the way. In general, I would like to thank the many undergraduate and graduate students who have been part of MaineSAIL (the Maine Software agents and AI Laboratory), as well as the researchers who have been affiliated with us. A special thanks goes to my current PhD student, Chris Wilson, not only for his work on CMBr, but also for many interesting and helpful

9 See the Wikipedia page "Neats and scruffies" for an overview of the long-standing dichotomy of approaches in AI.

discussions, and to the other members of the seabird image recognition team, Alex Revello, Logan Kline, Meredith Wilson, Matthew Brown and Cynthia Loftin. I would like to especially thank my late wife and colleague, Elise H. Turner, for her enormous contributions to the work over the years. I am also thankful to the various funding agencies who have supported this work, including the US National Science Foundation (BES–9696044, MRI 1919478) and the US Office of Naval Research (N0001–14–00–1–0–614, N0001–14–98–1–0648, N0001–14–96– 1–5009, N0001–14–96–1–5009), as well as the University of Maine AI Initiative Seed Grant program, most recently.

5.8. References

Aikins, J.S. (1980). Prototypes and production rules: A knowledge representation for computer consultations. PhD Thesis, Stanford University, CA.

Albert, E., Turner, E.H., Turner, R.M. (2007). Appropriate commitment planning for AUV control. *International Symposium on Unmanned Untethered Submersible Technology (UUST'07)*, Durham, NH.

Arritt, R.P. and Turner, R.M. (2003). Context-specific weights for a neural network. *Fourth International and Interdisciplinary Conference on Modeling and Using Context (CONTEXT'03)*, 29–39.

Baader, F. (2003). *The Description Logic Handbook: Theory, Implementation, and Applications*. Cambridge University Press, Cambridge.

Balsam, P.D. and Tomie, A. (eds) (1985). *Context and Learning*. Psychology Press, New York.

Berners-Lee, T., Hendler, J., Lassila, O. (2001). The semantic web. *Scientific American*, 284(5), 28–37.

Brézillon, P. (1990). Interpretation and rule packet in expert systems. Application to the SEPT expert system. *Proceedings of the International Conference on Knowledge-Based Computer Systems*, 78–87.

Brézillon, P. (ed.) (1993). Proceedings of the IJCAI-93 workshop on using knowledge in its context. Research report, LAFORIA, Université Paris VI.

Brézillon, P. (2003). Representation of procedures and practices in contextual graphs. *The Knowledge Engineering Review*, 18(02), 147–174.

Brézillon, P. (2014). A context-centered architecture for intelligent assistant systems. In *Innovations in Intelligent Machines – 4*, Faucher, C. and Jain, L. (eds). Springer, Cham.

Brézillon, P. and Abu-Hakima, S. (eds) (1995). *Working Notes of the 1995 IJCAI Workshop on Modelling Context in Knowledge Representation and Reasoning*. IJCAI, Montreal.

Brézillon, P. and Cavalcanti, M. (eds) (1997). *Proceedings of the International and Interdisciplinary Conference on Modeling and Using Context (CONTEXT-97)*. Rio de Janeiro, Brazil.

Brézillon, P. and Pomerol, J.-C. (1999). Contextual knowledge sharing and cooperation in intelligent assistant systems. *Le Travail Humain*, 62(3), 223–246.

Brézillon, P., Gentile, C., Saker, I., Secron, M. (1997). SART: A system for supporting operators with contextual knowledge. *International and Interdisciplinary Conference on Modeling and Using Context (CONTEXT–97)*, 209–222.

Chandrasekaran, B., Gomez, F., Mittal, S., Smith, J. (1979). An approach to medical diagnosis based on conceptual structures. *Sixth International Joint Conference on Artificial Intelligence*, Stanford, CA.

Cullingford, R.E. (1981). SAM. In *Inside Computer Understanding*, Schank, R.C. and Riesbeck, C.K. (eds). Lawrence Erlbaum Associates, Hillsdale, NJ.

Curtin, T.B., Bellingham, J.G., Catapovic, J., Webb, D. (1993). Autonomous oceanographic sampling networks. *Oceanography*, 6(3), 86–94.

Dey, A.K., Abowd, G.D., Salber, D. (2001). A conceptual framework and a toolkit for supporting the rapid prototyping of context-aware applications. *Human–Computer Interaction*, 16(2), 97–166.

Feltovich, P.J., Johnson, P.E., Moller, J.A., Swanson, D.B. (1984). LCS: The role and development of medical knowledge and diagnostic expertise. In *Readings in Medical Artificial Intelligence*, Clancey, W.J. and Shortliffe, E.H. (eds). Addison–Wesley Publishing Company, Reading, MA.

Fikes, R.E. and Nilsson, N.J. (1971). STRIPS: A new approach to the application of theorem proving to problem solving. *Artificial Intelligence*, 2, 189–208.

Furness, R.W. and Camphuysen, K. (1997). Seabirds as monitors of the marine environment. *IceS Journal of Marine Science*, 54(4), 726–737.

Ganesh, M.R., Blanchard, D., Corso, J.J., Sekeh, S.Y. (2020). Slimming neural networks using adaptive connectivity scores. arXiv preprint arXiv:2006.12463.

Giarratano, J.C. (1993). *CLIPS User's Guide.* NASA, Information Systems Directorate, Software Technology Branch, Lyndon B. Johnson Space Center, Houston, TX.

Giunchiglia, F. (1993). Contextual reasoning. *Epistemologia, Special Issue on I Linguaggie le Macchine*, 16, 345–364.

Gonzalez, A.J. and Brézillon, P. (2008). Comparing two context-driven approaches for representation of human tactical behavior. *Knowledge Engineering Review*, 23(3) 295–315.

Guha, R.V. (1992). Contexts: A formalization and some applications. Doctoral Dissertation, Stanford University, Department of Computer Science, CA.

Haarslev, V. and Möller, R. (2001). RACER system description. In *Automated Reasoning*, Goré, R., Leitsch, A., Nipkow, T. (eds). Springer, Berlin, Heidelberg.

Kellermann, K., Broetzmann, S., Lim, T.-S., Kitao, K. (1989). The conversation MOP: Scenes in the stream of discourse. *Discourse Processes*, 12(1), 27–62.

Kolodner, J.L. (1984). *Retrieval and Organizational Strategies in Conceptual Memory*. Lawrence Erlbaum Associates, Hillsdale, NJ.

Kolodner, J.L. (1993). *Case-Based Reasoning*. Morgan Kaufman, San Mateo, CA.

Kolodner, J.L., Simpson, R.L., Sycara-Cyranski, K. (1985). A process model of case-based reasoning in problem-solving. *International Joint Conference on Artificial Intelligence*, Los Angeles, CA.

Le Cun, Y., Boser, B., Denker, J., Henderson, D., Howard, R., Hubbard, W., Jackel, L. (1989). Backpropagation applied to handwritten Zip Code recognition. *Neural Computation*, 1(4), 541–551.

McCarthy, J. (1993). Notes on formalizing context. *IJCAI*, 555–560.

McGuinness, D. and Harmelen, F.V. (2004). OWL web ontology language overview. W3C technical report [Online]. Available at: www.w3.org/TR/owl-features.

Miller, R.A., Pople, H.E., Myers, J.D. (1982). INTERNIST–1, an experimental computer-based diagnostic consultant for general internal medicine. *New England Journal of Medicine*, 307, 468–476.

Noy, N.F., Crubézy, M., Fergerson, R.W., Knublauch, H., Tu, S.W., Vendetti, J., Musen, M.A. (2003). Protégé-2000: An open-source ontology-development and knowledge-acquisition environment. *American Medical Informatics Association (AMIA) Symposium*, 953–953.

Ramakrishnan, P. and Turner, R.M. (1997). Integrating partial-order planning into the Orca schema-based mission controller. *International Symposium on Unmanned Untethered Submersible Technology (UUST)*, Durham, NH.

Redmon, J., Divvala, S., Girshick, R., Farhadi, A. (2016). You only look once: Unified, real-time object detection. *IEEE Conference on Computer Vision and Pattern Recognition (CVPR)*, 779–788.

Rode, S. and Turner, R.M. (2015). Representing and communicating context in multiagent systems. In *Modeling and Using Context (Proceedings of the Ninth International and Interdisciplinary Conference on Modeling and Using Context)*, Christiansen, H., Stojanovic, I., Papadopoulos, G.A. (eds). Springer, Cham.

Schank, R.C. (1982). *Dynamic Memory*. Cambridge University Press, New York.

Shortliffe, E.H. (1976). *Computer-Based Medical Consultations: MYCIN*. Elsevier, New York.

Smith, R. (1980). The contract net protocol: High-level communication and control in a distributed problem solver. *IEEE Transactions on Computers*, C–29(12), 1104–1113.

Stein, G. and Gonzalez, A.J. (2014). Learning in context: Enhancing machine learning with context-based reasoning. *Applied Intelligence*, 41(3), 709–724.

Sticklen, J. (1987). MDX2: An integrated medical diagnostic system. PhD Thesis, Department of Computer and Information Science, The Ohio State University, OH.

Teege, G. (1994). Making the difference: A subtraction operation for description logics. *Knowledge Representation*, 94, 540–550.

Turner, R.M. (1989). Using schemas for diagnosis. *Computer Methods and Programs in Biomedicine*, 30(2/3), 199–208.

Turner, E.H. (1993). Organizing dialogue from an incoherent stream of goals. *Fifteenth International Conference on Computational Linguistics (COLING–92)*, 338–344, Nantes.

Turner, R.M. (1994). *Adaptive Reasoning for Real-World Problems: A Schema-Based Approach*. Lawrence Erlbaum Associates, Hillsdale, NJ.

Turner, R.M. and Turner, E.H. (1998). Organization and reorganization of autonomous oceanographic sampling networks. *1998 IEEE International Conference on Robotics and Automation (ICRA'98)*, 2060–2067, Leuven.

Turner, R.M. and Turner, E.H. (2001). A two-level, protocol-based approach to controlling autonomous oceanographic sampling networks. *IEEE Journal of Oceanic Engineering*, 26(4), 654–665.

Turner, R.M. and Turner, E.H. (2004). Self-organization and reorganization of multi-AUV systems: CoDA project overview. *IEEE/OES Autonomous Underwater Vehicle Workshop on Multiple AUV Operations (AUV'04)*, Sebasco Estates, ME.

Turner, E.H., Chappell, S.G., Valcourt, S.A., Dempsey, M.J. (1994). COLA: A language to support communication between multiple cooperating vehicles. *1994 Symposium on Autonomous Underwater Vehicle Technology (AUV'94)*, 309–316.

Turner, E.H., Turner, R.M., Albert, E. (2005). Placing newly-arising goals in the proper context. *Notes for the Workshop on Cooperative Systems and Context, Fifth International and Interdisciplinary Conference on Modeling and Using Context (CONTEXT'05)*, Paris.

Turner, R.M., Rode, S., Gagne, D. (2013). Distributed, context-based organization and reorganization of multi-AUV systems. *Journal of Unmanned System Technology (JUST)*, 2(1), 1–9.

Whitsel, L. and Turner, R.M. (2015). Using contextual knowledge for trust strategy selection. In *Modeling and Using Context (Proceedings of the Ninth International and Interdisciplinary Conference on Modeling and Using Context)*, Christiansen, H., Stojanovic, I., Papadopoulos, G.A. (eds). Springer, Cham.

Wilson, C. and Turner, R.M. (2015). Modeling erroneous human behavior: A context-driven approach. In *Modeling and Using Context (Proceedings of the Ninth International and Interdisciplinary Conference on Modeling and Using Context)*, Christiansen, H., Stojanovic, I., Papadopoulos, G.A. (eds). Springer, Cham.

Wilson, C. and Turner, R.M. (2017). Context and the virtual human. In *Modeling and Using Context (Proceedings of the Tenth International and Interdisciplinary Conference on Modeling and Using Context)*, Brézillon, P., Turner, R.M., Penco, C. (eds). Springer, Berlin.

Zadeh, L.A. (1979). A theory of approximate reasoning. In *Machine Intelligence*, Hayes, J., Michie, D., Mikulich, L. (eds). Halstead Press, New York.

6

Using Context to Help Identify the Emotional State of a Human in a Conversation

This chapter describes how the use of context can help identify the emotion being expressed by a person when they are uttering sounds paralinguistically – that is, vocalizing sounds without understandable words. Paralinguistic affect recognition is an active area of research that has seen mixed success in automatically inferring the emotion of an individual from such sounds alone. The current research has centered on the use of classifiers that use as inputs a variety of features extracted from the vocalized sound to select the emotion most probably being expressed by the human. The objective of the research reported here is to determine whether the use of context can help in making such inferences. We use techniques adopted from the field of sentiment analysis to model the context being experienced by the actor vocalizing the sounds to further filter an initial list of possible emotions proposed by a conventional classifier-based approach. We found that mapping keywords selected from a textual description of the actor's context to emotions defined by valence and arousal coordinates, as contained in a sentiment analysis library, can improve the selection accuracy by approximately 20% over the published state of the art for 17 possible emotions. This chapter describes the context modeling process used and the overall system designed to do the affect recognition.

6.1. Introduction and background

It has been widely reported in the literature that context is an integral part of our cognitive process. See Hollister et al. (2019) for an in-depth discussion of this topic,

---

Chapter written by Andreas H. MARPAUNG and Avelino J. GONZALEZ.

including an overview of the work of others on this subject. Less frequently discussed is the influence a context has on the emotions that humans can feel when they are in that context. For example, we would generally expect attendees at a funeral service to feel sadness; those at a theme park or a party to feel joy; those about to undergo heart surgery to feel a twinge of anxiety. While these examples do not reflect absolute mappings (e.g. we could feel joy at the funeral of a sworn enemy), they can provide a strong indication of what emotions the involved persons would be most likely to feel. So, certain situations (i.e. contexts) can usually be associated with certain emotions.

The research that we discuss here focuses on identifying the emotions felt by a human speaker whose situation is known and described in text (albeit briefly – in only two or three sentences at most). However, any words spoken by the speaker are not understood by the listener (assume a language totally foreign to the listener, or a distance too great to make out the words spoken). Nor does the listener have any visual contact with the speaker, precluding the interpretation of facial expressions or body language. However, the listener can hear the (unintelligible) sounds of the speech uttered by the speaker and is aware of their environment and possibly their purpose for being in that environment. This is called *paralinguistic affect recognition*.

Paralinguistic sounds do not give us much to go by, as shown by the relatively weak results achieved in identifying the emotions behind the paralinguistic utterances through various classifiers. We discuss this is in section 6.3. So, our work sought to add the element of context that, when available, can hypothetically help in paralinguistic affect recognition. In section 6.2, we set out our hypothesis and the use case; in section 6.3, we discuss the works of other authors who have done research in this same vein, even if not exactly for the same purposes. Section 6.4 includes a discussion of how *sentiment analysis* can be used in context modeling and its relevance to our work. These sections are followed by an extensive discussion in section 6.5 about our method to identify the context and propose the most likely emotion, and in section 6.6, we provide one example of how our approach works. Finally, section 6.7 summarizes and makes conclusions about our work, including a discussion of what we believe to be the most important future research.

6.2. Use case and research hypothesis

As mentioned, our overall research comes under the paralinguistic affect recognition umbrella. Identifying a speaker's emotional state from paralinguistic signals could help when speaking to someone in a (non-video) telephone conversation, in a language that the listener does not understand well or through a very noisy communication medium that results in very garbled language. We are then left with having to identify a speaker's emotions paralinguistically and contextually.

Therefore, our hypothesis is that knowing the context in which that person finds themselves can be helpful in identifying their emotional state when we otherwise only have paralinguistic data.

Our wider research employs classification systems to initially infer the emotion of the speaker based only on the sounds of their utterances. The output of the classifiers is a list of possible emotions, listed in decreasing order of probability. Then, our context modeling work identifies and integrates the context to disambiguate the multiple emotions proposed by the classifiers to arrive at the one emotion most likely to have been exhibited by the speaker. We label those two separate and sequential steps as phase I and phase II; our focus in this chapter is on phase II. Phase I is the context-free phase where the speech audio is segmented into several smaller windows and their acoustic features are extracted within these windows. We experimented with several classification algorithms to classify the sounds uttered by the speaker to a set of possible emotions. Phase II takes the top X of these emotions according to their probability of correctness by the classifiers from phase I and applies the contextual information to more accurately classify the sound uttered by the speaker being the one most likely emotion to be expressed by the speaker.

We have built the *context-based paralinguistic affect recognition system* (CxBPARS) which incorporates phases I and II to improve the accuracy of the results over those reported in the literature that only use the classifiers (the equivalent of phase I). We have reported on the CxBPARS elsewhere, and we refer interested readers to Marpaung (2019) for a detailed and full description of it. Therefore, in this chapter, we focus on and describe in greater detail how we identify and model the context in which the speaker finds themselves and how we link that context to possible speaker emotions.

Our approach works by first selecting keywords from a brief description of the speaker's situation (of course this is assuming such a description exists, which is not always the case), processing these to discover the context and then accessing a context database that links the context discovered to a (short) list of possible emotions that could be felt by the speaker when in that context. This context database currently has to be built manually, but we hope in the future to at least partially automate that process to lessen the burden for whoever is to create such a database.

Finally, we should mention that unlike all other works published in the current literature, discussed below, our research includes 17 emotions. These are admiration, amusement, anger, anxiety, contempt, despair, disgust, fear, interest, irritation, joy, pleasure, pride, relief, sadness, surprise and tenderness. All other published research works only include 2, 6 or at most 12 emotions from which to choose. This inherently makes our problem significantly more difficult.

6.3. Related works

The interest in measuring the relationship between words and emotions began to grow in the early 1990s. One of the earliest reported works was the Geneva Expert System on Emotion, or GENESE (Scherer 1993) which used Euclidian distance measures to predict the emotion that a user experienced. GENESE was developed as an attempt to formalize the Appraisal Theories of Emotion (Scherer 1993, 1999).

The premise of the appraisal theory of emotion is that emotions are differentiated and elicited based on a person's appraisals or subjective evaluation, which in turn are based on a set of dimensions or criteria (Scherer 1993, 1999). To detect and assess the significance of an organism's environment, the appraisal theories specify the appraisal criteria (variables); some of these variables include coping potential or control, certainty, goal relevance and goal congruence (which refer to the relevance and congruence of events for the goals) and agency (an event caused by oneself, by others or by impersonal circumstance). For example, an individual walking in a jungle encounters a sleeping lion (agency) that can kill them if it is awakened (goal congruence). The individual is not sure (certainty) whether they can leave the area quietly without waking up the lion (control). Under these circumstances, the individual will very likely experience fear.

GENESE studies individual differences in appraising different variables of stimulus evaluation checks (SECs). Despite these differences, each emotion has some common patterns. In their study, to predict the emotion felt by a test subject, the individual must think of a situation in which they experienced a strong emotion. Then, the individual has to answer a set of 15 questions related to different variables of SECs for a particular situation about which they were thinking. Based on these responses from a subject, the GENESE program predicts the test subject's emotion. If the first predicted emotion is not correct, after receiving confirmation from the test subject, the second smallest distance is then offered. Based on testing with 200 reported emotion situations, the system could predict 78% of situations correctly after the first two times.

The computational model of appraisal theory of emotion through GENESE sparked other research that employed expert systems and computer modeling to propose other computational models of emotions. This body of research included the work by Chwelos and Oatley (1994) and Wehrle and Scherer (1995). Their works served as inspiration for our approach to linking keywords to emotions. These works derived the emotional connotation of words directly. However, we take the extra step of inferring the context in which the agent finds themselves/itself to avoid misleading statements that could skew the inference. For example, in a statement such as, “In spite of my pain, I dream of being at Disney World”, the emotion of the speaker could be construed as joy if only “Disney World” is chosen as the keyword.

Inspired by the increasing attention toward the field of sentiment analysis, we have integrated some concepts of sentiment analysis into our context-based approach. Sentiment analysis (SA), also known as *opinion mining* or *subjectivity analysis*, operates at the intersection of artificial intelligence, natural language processing and information retrieval. SA focuses on determining the sentiment expressed in text. This field of research studies the sentiment, the phenomena of opinion, attitude, evaluation, emotion and appraisal (Liu 2012).

Since its inception in 1990, the research interest in the sentiment analysis through text domain had seen increasing growth. In 1992, Hearst proposed the automatic detection of directionality as a measurement of subjective states (Hearst 1992). Most of the notable works in this domain have been previously reported in Mohammad et al. (2015).

Perez-Rosas et al. (2017) built a sentiment analysis classifier to identify the polarity of spoken opinions in cellular phones and fiction books. Through their study, three research questions were answered. First, through an SVM classifier[1], they were able to show that the automatic sentiment classifier could rely on the linguistic component of online spoken reviews. Second, they found that manual transcription outperformed automatic transcription in terms of accuracy. Third, they found that the information orally encoded in multimodal reviews was more informative than in written reviews as people tend to use additional resources such as body gestures and intonations to help them deliver messages more accurately.

Pereira et al. (2014) proposed an algorithm to retrieve the repository of documents based on the sentiment polarity criteria expressed in natural spontaneous speech. Furthermore, in a more challenging setting, as a result of speech spontaneity, unknown background environments, variable source and channel characteristics, accents and topic diversity, Kaushik et al. (2017) proposed an algorithm to extract speaker sentiment from natural audio streams in YouTube.

With the influence of social media and widely available complementary data streams such as YouTube, the interest of researchers to work on the multimodal sentiment analysis has steadily grown as it offers promising means to analyze vocal and facial expressions in addition to the textual content (Soleymani et al. 2017).

---

1 SVM (support vector machine) is a supervised learning model to analyze data for classification, regression analysis or outlier detection (Cortes and Vapnik 1995). The algorithm maps training points to higher-dimensional space to maximize the gap between the two categories.

With respect to affect recognition, interest in this area through different modalities gained more attention after the birth of affective computing (Picard 1997). These modalities include vision (Yu et al. 2018; Zhang et al. 2018; Zeng et al. 2018), body gestures (Kaza et al. 2016; Noroozi et al. 2018), speech (Yadollahi et al. 2017; Sailunaz et al. 2018) and multimodal approaches (Zhang et al. 2016, 2018; Huang et al. 2019), where more than one of these modalities is used in conjunction. Deep neural networks have gained popularity in the speech paralinguistic domain. The first documented approach in affect recognition with speech affect recognition dates back 14 years (as per Schuller et al. (2018)) when Wollmer et al. (2008) used three-layered long–short-term memory (LSTM) recurrent neural networks (RNNs)[2]. Three years later, Stuhlsatz et al. (2011) proposed the next deep neural network approach where they used a generalized discriminant analysis (GerDA) to learn discriminative features of the raw signal and restricted Boltzmann machines (RBM) to pre-optimize the network parameters. Five years later, Trigeorgis et al. (2016) proposed the first true end-to-end computational paralinguistic system using convolution recurrent neural network architecture for continuous affect recognition (Schuller et al. 2018). For the past decade, many works using deep neural networks have been proposed. Yet, surprisingly, context has rarely been one of the modalities considered.

Some recent speech paralinguistic works include Meng et al. (2019) and Zhao et al. (2019). Zhao et al. used two databases: 1) Berlin EmoDB corpus, which focused on seven emotions: anger, boredom, disgust, fear, happy, neutral and sadness, and 2) IEMOCAP database, which focused on six emotions: anger, excitement, frustration, happy, neutral and sadness. Their approach achieved recognition accuracies of 95.3% (for speaker-dependent experiments) and 95.9% (for speaker-independent experiments) for Berlin EmoDB and 89.2% (for speaker-dependent experiments) and 52.1% (for speaker-independent experiments) for IEMOCAP database. Meng et al. achieved recognition accuracies of 90.4% (for speaker-dependent experiments) and 84.9% (for speaker-independent experiments) for the Berlin EmoDB and 74.9% (for speaker-dependent experiments) and 69.3% (for speaker-independent experiments) for the IEMOCAP database.

6.4. Sentiment analysis as a way to model context

Of the research works mentioned in section 6.3 about sentiment analysis, we borrow one of them for our own work – the Sentiment Analysis Dictionary. Such dictionaries are currently being used in four lines of research: (1) research on how

---

2 LSTM RNN networks are excellent for classifying, processing and making predictions based on time series data because there can be lags of unknown duration between important events in a time series.

emotions are produced and perceived and their consequences on human behavior (Verona et al. 2012); (2) research on processing and memorizing words (Kousta et al. 2011); (3) research on text-based sentiment analysis (Leveau et al. 2012) and (4) research on the emotional values of new words by comparing them to validated words (Bestgen and Vincze 2012). Verona et al. (2012) studied the responses between those participants without a personality disorder, and those participants with an antisocial personality disorder (APD) when they were exposed to neutral and negative words. From the study, they found that the APD group showed greater processing of negative emotional words versus participants without a personality disorder who showed greater processing of the neutral words. Kousta et al. (2009) studied the impact of positive, negative and neutral words in a lexical decision experiment and found that the participants responded faster to the positive and negative words. Leveau et al. (2012) estimated the degree of valence and arousal evoked by texts on the basis of word measures. Bestgen and Vincze (2012) computed the affective values of 17,350 new words from known semantically related words.

Bestgen and Vincze (2012) was one of the works in the computational linguistics domain that automatically tagged the affective valence (also known as *affective polarity* or *semantic orientation*) to unlabeled words from some emotional lexicons. Valence indicates how strongly a positive or a negative connotation a certain word has (Bestgen and Vincze 2012). For the past several decades, most sentiment analysis researchers have used the ANEW norms database (Bradley and Lang 1999), which has 1,034 words as its base emotional lexicon and have expanded their lexicons using some computational methods.

Other researchers have also used WordNet (Miller 1995) as their starting point; these include Esuli and Sebastiani (2006), who estimated the similarity between words from the relation of synonymy using a bootstrapping algorithm. Esuli and Sebastiani (2006) developed the SentiWordNet using a semi-supervised learning procedure[3] to assign the valence to each WordNet synset. Synset instances are defined as groupings of synonymous words in the same concept.

In addition to several emotion lexicons that were generated through some machine learning techniques, some researchers also attempted to expand their word collections in their proposed lexicons obtained through a crowd sourcing method; Kuperman et al. (2014) is an example of lexicons collected through crowd sourcing. We use this corpus for our work. The next section further discusses this corpus.

3 Semi-supervised learning is a machine learning approach that combines a small amount of labeled data (in supervised learning) with a large amount of unlabeled data (in unsupervised learning) during training.

6.5. Our approach to the problem

Our work makes use of the emotional implications of words and links these words to the contexts being experienced by the actors who utter the sounds used to identify their emotions in real time. Our contexts are modeled by the keywords used to identify their situation and what emotions a person in that context is likely to experience. We explain this in greater detail in this section, where we first briefly discuss our general approach to what we now refer to as *context-based paralinguistic affect recognition* before we dive into a description of our approach to discovering the context and how to use it to disambiguate the several potential emotions proposed by phase I.

6.5.1. *Our overall approach to paralinguistic affect recognition*

The overall architecture of CxBPARS is shown in Figure 6.1. Our work was originally inspired by a human study (Marpaung and Gonzalez 2017, 2021) that confirmed that knowing the context of the speaker indeed plays a significant role in the ability of a (human) observer (actually, a listener) to identify the emotion being exhibited by a speaker. As we mentioned earlier, CxBPARS has two phases: (I) a context-free phase to initially propose a set of emotions through classification of the sounds uttered by the speaker, and (II) a context-centric phase designed to improve classification accuracy by disambiguating the output of the first phase. The context-free phase includes data segmentation, feature extraction and classification systems. The inputs to the context-free approach of phase I are the speech audio captured by a microphone and recorded. Then, this speech audio is segmented into several smaller windows and their acoustics features are extracted within these smaller windows. For phase I, we experimented with several classifiers, for example, SVM, AdaBoost[4], Ensemble[5] and deep learning neural networks[6]. We found that AdaBoost gave the best classification result empirically. The classification results are a set of possible emotions whose acoustics in some ways would normally match the sounds uttered by the speaker. Each of these is assigned a

4 AdaBoost (Adaptive Boosting) is a statistical classification meta-algorithm that combines the output of the weak learners' algorithms with the weighted sum that represents the final output of the boosted classifier (Schapire and Freund 2012).

5 Ensemble uses multiple learning algorithms to obtain better predictive performance than the ones obtained from any of the constituent learning algorithms (Opitz and Maclin 1999).

6 Deep learning neural networks is a machine learning approach that mimics the human brain where it takes information and generates an output based on their knowledge through adaptation and learning, without having to be reprogrammed (Schmidhuber 2015).

probability of correctness by the classifiers and they are then fed as inputs to the context modeling module (phase II)

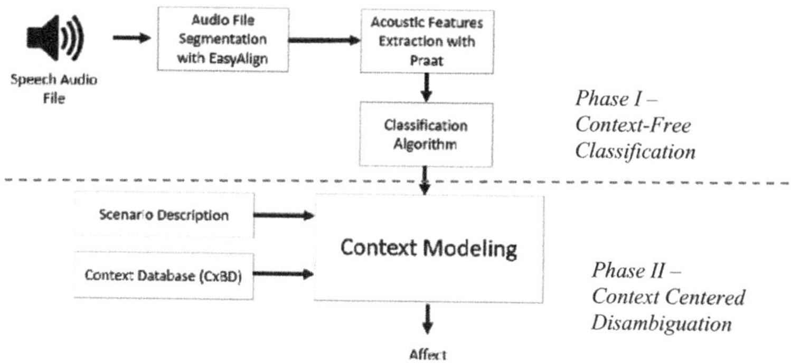

Figure 6.1. *CxBPARS architecture*

The context-centric phase II also requires a brief textual description of the scenario, a sentiment analysis dictionary and a context database. Its output is the one emotion most likely to be expressed by the speaker from the initial set generated by the first phase. This chapter only describes phase II in detail.

6.5.2. *A (very) brief description of phase I (context-free classification)*

Figure 6.2 shows the flow diagram of what happens in phase I of this process. The details about which classifiers were used, how they were used, with what data they were trained and how the features were extracted from the audio file data provided for our testing are covered extensively in Marpaung (2019). Space limitations and the tangential nature of the first phase to the subject of context of this book preclude its full inclusion here.

Output of phase I: The top X output tuples: $\{(A_{top\_1}, p_{top\_1}), (A_{top\_2}, p_{top\_2}) \ldots (A_{top\_x}, p_{top\_x})\}$, where A_{top_1} represents the affect (emotion) that the speaker tries to convey and p_{top_1} represents its probability; X represents an integer between 1 (only the top emotion is passed to phase II) and 17 (all emotions are passed to phase II). The value of X is set by the user or the developer. Of course, if X = 1, there would be no need for phase II, and if X = 17, there would be no need for phase I. Practically speaking, X is ideally in the 9 to 11 range as we previously determined through trial and error. See Marpaung (2019) for details on this selection process.

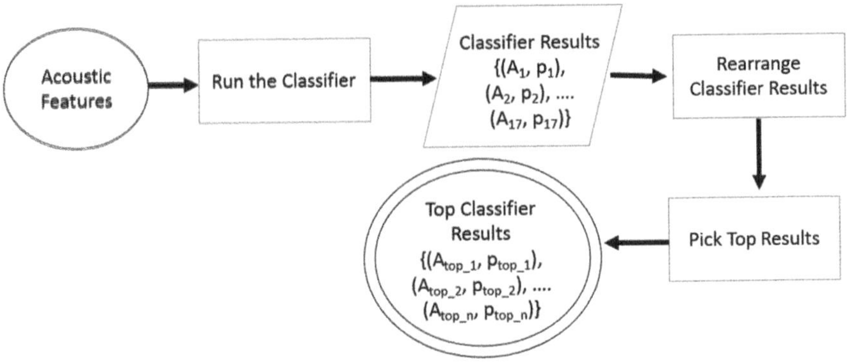

Figure 6.2. *Flow diagram of the phase I classification process*

6.5.3. *Phase II – the context-centered process*

Before describing the phase II process, a description of the data used for training the classifiers of phase I as well as for the final summative testing of our prototype system needs to be provided in order to gain a better understanding of the phase II process.

6.5.3.1. *The GEMEP database*

We used the *Geneva Multi-modal Emotion Portrayals* (GEMEP) database (Banziger and Scherer 2010) in our research. We used it in our study because it is the only publicly available data that contains a description of the (imaginary) situation the speaker was in when their utterances were recorded. Therefore, GEMEP contains the two input components that are meaningful to our study: 1) the audio files containing recordings of meaningless, pseudo-speech sentences uttered by actors acting out a prescribed emotion, and 2) the aforementioned accompanying brief textual description of the imaginary situation faced by the actors who uttered the sounds recorded in each of the audio files. The former provides the inputs for phase I, while the latter allows our system to construct the context model for phase II. Our work includes all 17 emotions contained in the GEMEP corpus, as described in section 6.2.

The GEMEP audio files were generated by 10 (five male and five female) professional French-speaking actors who had previously worked in artistic productions (age range: 25–57 years old; mean age: 37.1 years). These actors portrayed several affective states by uttering one of these two meaningless pseudo-speech sentences: "nekal ibam soud molen!" (intoned as a declarative statement or exclamation) or "koun se mina lod belam" (intoned as a question) for each affective state. The actors uttered these pseudo-speech sentences in a way that exhibited the emotion to be expressed for that audio file.

With the exception of disgust and surprise, each emotion had 10 audio files, spoken one each by the five different male actors and the five different female actresses; disgust only had four audio files, produced by two male and two female actresses, while surprise had five files, produced by three male and two female actresses.

To induce and elicit certain emotions, the actors used the *felt enacting technique*, which is typically used in theatrical settings. With this technique, an actor recalls or has mental imagery of a relevant personal experience to help them enact certain emotions based on the written script given to them prior to the recordings. These original scripts describe the context of the emotions to be felt by the actor and are therefore important to our research. They are written in French. For our study, we translated the original scripts from French to English using Google Translate (Google 2022). An example of a scenario description in French along with its English translation follows.

> **French:** "J'ai gagné une somme fabuleuse au loto. Je ne m'y attendais pas du tout, j'avais acheté un billet par hasard en allant boire un café avec un ami. Je suis fou de joie lorsque je le découvre et je cours l'annoncer à mon/ma partenaire (mes parents, mes enfants)." (Banziger and Scherer 2010)

> **English:** "I won a fabulous lottery prize. I did not expect it at all, I had bought a ticket by chance while going for a coffee with a friend. I am overjoyed when I find out and I run to tell my partner (my parents, my children)." (Google 2022)

To build the context-centric system, we used two keywords extracted from these GEMEP scenario descriptions as described below.

6.5.3.2. *The phase II process*

In its most basic form, the general plan of the phase II process is to:

1) extract keywords from the textual description of the situation;

2) use these keywords to identify the context in which the speaker finds themselves;

3) use techniques from sentiment analysis to disambiguate the possible emotions.

First, we discuss general descriptions of each of these general steps. Then, we describe these three steps in detail, starting with keyword extraction. Then, we provide an algorithm that sets the process into effect and that provides yet more details about how phase II works. We start with the general descriptions.

Step 1 - Keyword extraction: this is a comparatively simple step where two keywords are selected at random from the two- or three-sentence scenario description provided by GEMEP. Currently, our approach extracts two verbs because our prior human subject study (Marpaung and Gonzalez 2014, 2021) indicated that an action context (i.e. a verb) was more useful than a relationship context (more likely described with nouns).

However, in a real-world application, there would be no GEMEP description, so other sources of information to identify the current context of the speaker must be found. One such source could be a person's electronic calendar, possibly in their smart phone, where brief descriptions of meetings, travels or other such experiences may be kept along with the time and place for such like. These time-stamped descriptions may provide indications of the person's context. So, an example of keyword extraction from such a calendar app in a subject's smart phone would have the following entry:

Sep 24, 2019, 11 a.m. to Noon – *Visit dentist to extract tooth.*

From this entry, the keywords (i.e. verbs) extracted would be *visit* and *extract*.

In other cases, the context may arise suddenly and unexpectedly (e.g. a traffic accident or a sudden illness) for which there may be no prior documentation. In such cases, it may be possible that a limited set of words spoken by the speaker may be captured and understood (assuming a common language), which could then serve as keywords. We leave this case for future research.

We do realize that by limiting the keywords to action words, we may fail to capture a scenario's important essence. In this same example, such would not capture a more important keyword – "dentist" – which holds a more meaningful implication of this visit when compared to simply saying "visit". Therefore, in our future research, we plan to experiment with different ways to extract keywords that are more meaningful, such as using natural language understanding techniques to tie multiple keywords into phrases, such as "visit dentist", which would be significantly more meaningful than the keywords selected separately, as done in this work.

Step 2 - Linking keywords to contexts: now that it has identified the keywords, the phase II process next seeks to link them to contexts. For this work, we introduce the *context database* (CxDB). In real life, this database could be likened to an individual's recollections of familiar contexts associated with everyday life, so it can be said to reflect human experiences. For example, when we go to a library (say, the

LIBRARY context), we have an image associated with this context of *quiet_place*, *books*, *shelves*, *librarians* and *tables*. Likewise, we may feel certain emotions when we make a visit to a library, such as amusement, interest, joy, pleasure, anxiety, fear, irritation and possibly surprise. Each piece of contextual knowledge in the CxDB has the following attributes:

– context name;

– keywords associated with this contextual knowledge;

– possible emotions that can be experienced by someone in this context.

An example of contextual knowledge is the following:

– **context**: library;

– **keywords**: quiet_place, books, shelves, librarians, tables, computers, document databases, read, research;

– **emotions**: amusement, interest, joy, pleasure, anxiety, fear, irritation, surprise.

Of course, not all emotions defined in a context are applicable to a specific instance of that context. The appropriate emotions within a context will depend on other issues, which we propose to identify indirectly through the paralinguistic aspects of speech of phase I. A search and match process will identify the contexts in the context database that match at least one of the keywords extracted.

As our initial approach, as there is no available pre-existing context database that fits into our affect recognition puzzle, we manually built a CxDB containing 50 contexts. We fully realize that context alone, even if it could be defined at a high granularity, can neither uniquely nor always identify the emotions felt by an individual. Furthermore, defining a context at the desired granularity may be difficult to do simply from a handful of keywords. Our goal in using the context database is to narrow down the emotions possibly expressed by the speaker in the identified context from the top X emotions of interest passed on by the phase I classifiers to some lower number – hopefully, one. We hypothesize that our algorithm will produce more accurate results for different scenarios. Machine learning methods may be able to automate the creation of this database, but this is also left for future research.

Step 3 - Word-affect mapping: this step is the most important and complex of the three, and therefore requires the most space to describe and explain it appropriately. Inspired by reported research works in the field of sentiment analysis that we discussed in section 6.4, we now provide more detail about how we used this concept to model our contexts.

We briefly discussed Kuperman et al.'s (2014) emotion lexicon at the end of section 6.4. They report an emotion lexicon with a collection of 13,915 affective words (63.5% nouns, 12.6% verbs, 22.5% adjectives and 1.4% unspecified parts of speech). Kuperman et al.'s lexicon is an example of a Sentiment Analysis Dictionary. In comparison, Bestgen and Vincze (2012) had 17,350 words in their emotion lexicons that were mapped in the *semantic space* (SS). The semantic space in the natural language domain is defined as the space to represent natural language to capture some meanings (Baroni and Lenci 2010). The SS stems from two core challenges in the natural language domain: (1) vocabulary mismatch – some meanings can be expressed in many ways, and (2) ambiguity of natural language – the same term can have several meanings.

The English subsidiary theorems by Kuperman et al. (2014) were built using participants who self-identified as US residents, recruited via Amazon's Mechanical Turk crowd sourcing website. The goal of this study was to investigate how people respond emotionally to different types of words. Three types of metrics were measured:

a) **valence** – measures how pleasant the stimulus is with a scale that ranges from 1 (unpleasant (arousal: excited, dominance: controlled)) to 9 (pleasant (arousal: excited; dominance: controlled));

b) **arousal** – measures how high/low the intensity of the emotion is (excitement) provoked by the stimulus with a scale that ranges from 1 (calm (valence: pleasant, dominance: controlled)) to 9 (excited (valence: pleasant, dominance: controlled));

c) **dominance** – measures how much control is exerted by the stimulus with a scale that ranges from 1 (in control (valence: pleasant, arousal: excited)) to 9 (controlled (valence: pleasant, arousal: excited)).

Participants spent approximately 15 minutes (each session) to rate around 350 words. Post-session, some information, for example, age, gender, first language(s), country/state resided in most between birth and age 7 and education level, were collected. Kuperman et al. reported their studies on the relationships between demographics and ratings, between gender differences and ratings, and semantic categories. We used this dictionary because it is not domain-specific, which is different from the other lexicons. This emotion lexicon also provides the largest collection of words that covers primary demographic information, such as age, gender and education.

To build a contextual model, we used Kupperman et al.'s sentiment analysis dictionary that captures the emotional dimensions (ED) of words – the valence (positive or negative), the arousal (high or low) and the emotion association lexicons. Examples of valence and arousal association lexicons include: (1) scream – negative emotion (scream is usually associated with high arousal and negative valence), (2) joy – positive emotion (joy is usually associated with high arousal and positive valence) and (3) a chair – neutral (a chair is not associated with positive or negative valence or high or low arousal). Examples of affect association lexicons include: (1) shout – anger (shout is associated with anger), (2) laughter – happiness (laughter is associated with happiness) and (3) furniture – neutral (furniture is not associated with any of the emotions).

We created a semantic space coordinate system as a two-dimensional Cartesian plane with the following two axes:

1) the horizontal valence axis (or V-axis) with the valence coordinate (VC) values ranging from 0 to 9, where 0 represents unpleasantness and 9 represents pleasantness;

2) the vertical arousal axis (or A-axis) with the arousal coordinate (AC) values ranging from 0 to 9, where 0 represents "uncalmness" and 9 represents calmness.

The values for valence, arousal and dominance are provided in the emotion lexicons of Kuperman et al. (2014). However, because most emotion theories in psychology correlate affect with valence and arousal only, our work focuses only on valence and arousal and neglects the dominance factor. Our contextual modeling begins with the mapping of all keywords extracted from the description of the situation, with their affective wording association (their values of valence and arousal) from Kuperman et al.'s dictionary. In our work, the keyword coordinate pair is referred to as the *keyword vector* consisting of the keywords VC and AC. For example, lottery (VC = 6.33, AC = 6.05), win (VC = 6.97, AC = 5.61) and receive (VC = 7.14, AC = 4.3). Figure 6.3 shows these three keyword vectors in the SS. Figure 6.3 shows the words lottery (VC = 6.33, AC = 6.05), win (VC = 6.97, AC = 5.61), receive (VC = 7.14, AC = 4.3) and gift (VC = 7.27, AC = 4.64) in the SS.

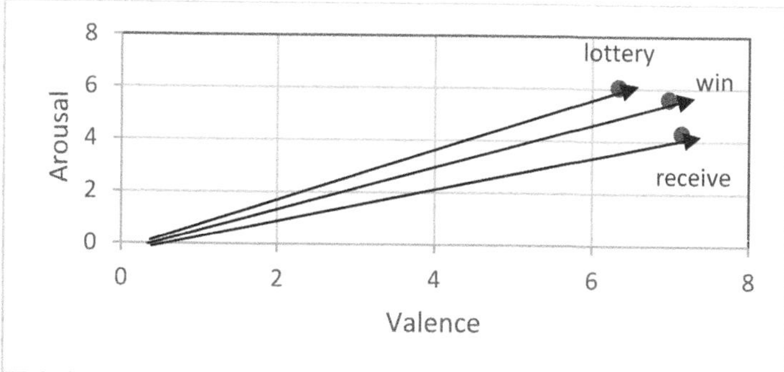

Figure 6.3. *Three keywords in the semantic space*

Emotion	Abbreviation	Valence	ED - Valence	Arousal	ED - Arousal
Admiration	ADM	7.58	Positive	5.52	Low
Amusement	AMU	7	Positive	4.82	High
Anger	ANG	2.5	Negative	5.93	High
Anxiety	ANX	2.38	Negative	4.78	Low
Contempt	CON	3.22	Negative	4.61	Low
Despair	DES	3.61	Negative	4.64	Low
Disgust	DIS	3.32	Negative	5	Low
Fear	FEA	2.93	Negative	6.14	High
Interest	INT	6.66	Positive	4.36	Low
Irritation	IRR	3.38	Negative	5.5	Low
Joy	JOY	8.21	Positive	5.55	High
Pleasure	PLE	7.8	Positive	6.8	Low
Pride	PRI	6.5	Positive	5.54	High
Relief	REL	6.63	Positive	4.42	Low
Sadness	SAD	2.4	Negative	2.81	Low
Surprise	SUR	7.44	Positive	6.57	High
Tenderness	TEN	6.89	Positive	3.1	Low

Table 6.1. *Seventeen emotion vectors (Kuperman et al. (2014) for valence and arousal coordinates, and Banziger and Scherer (2011) for ED values)*

For the initial step, we mapped each of the 17 emotions into the same SS. The VC and AC coordinates for these emotions were likewise taken from Kuperman et al.'s dictionary. For example, joy can be defined as (VC = 8.21, AC = 5.55) and sadness as (VC = 2.4, AC = 2.81). This is shown in Table 6.1, which lists all the *affect vectors* and the arousal and valence emotional dimensions (Banziger and Scherer 2010). This table shows, for example, that "anger" has a negative valence and a high arousal property; joy has a positive valence and a high arousal property; and sadness has a negative valence and a low arousal property. Figure 6.4 shows all affects plotted in the SS.

Next, we describe how we map the keywords, which we obtain from the scenario description, into the same SS coordinate system. By mapping the emotions and the keywords, we can calculate the distances between each keyword to all 17 emotions and associate the keyword of interest to the affect with the shortest distance.

The keywords are mapped into the SS in the same fashion as the emotions. The VC and AC values for each keyword vector are obtained from Kuperman et al.'s dictionary.

Finally, to identify the emotion most closely associated with each keyword, the Euclidian distance is measured between each keyword and each emotion's coordinates. These emotions are the ones that survive the filtering of the phase I classifiers and the CxDB. For example, let us say that the emotions pride, relief, interest and amusement all survive the aforementioned filtering. Then, calculation of Euclidian distances between the keyword "lottery" to these emotions are: (1) to pride: 0.54; (2) to relief: 1.66; (3) to interest: 1.72 and (4) to amusement: 1.40. Once calculated, the emotion with the shortest distance to the keyword of interest is associated with this keyword. Based on this calculation, and as shown in Figure 6.5, the keyword "lottery" is best associated with pride. Likewise, from Figure 6.5, it is shown graphically that the keywords "win", "gift" and "receive" are best associated with amusement.

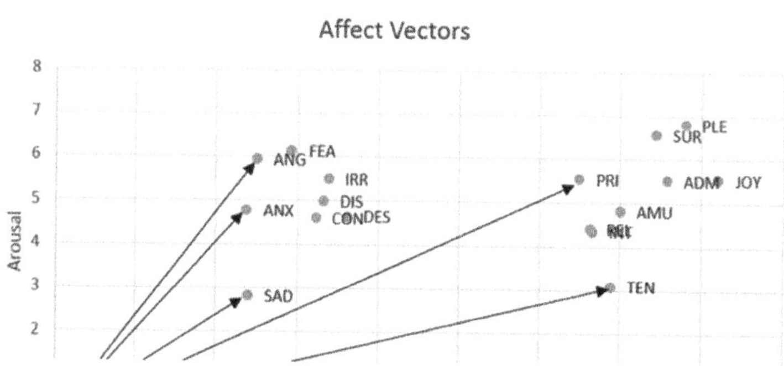

Figure 6.4. *Affect vectors in the SS (the arrows represent the vectors) – only some of the arrows are shown*

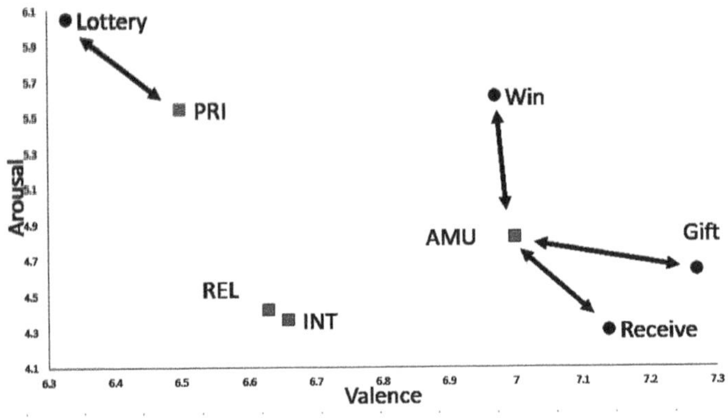

Figure 6.5. *Word affect mapping. For a color version of this figure, see www.iste.co.uk/brezillon/context.zip*

6.5.3.3. *Putting it all together*

The phase II process follows the following procedure – a more granular version of the general three-step model process above.

1) Extract the keywords KW1 and KW2 from a scenario description, from GEMEP, a calendar app or other such source.

2) Compare the keywords KW1 and KW2 with keywords from each context entry in the context database (CxDB) to produce the *matched context list* (CxREG).

3) For each matched context in CxREG, extract from each of these contexts the emotions associated with it and name this group CxEG (*context emotion group*).

4) Eliminate any duplicates in CxEG.

5) Find the intersection between CxEG and the list of emotions produced by the phase I classifiers (called the *classifier emotion group* or *CEG*). This intersection is called the IEG (*intersection emotion group*). For the purpose of consistency, we place the top X classifier emotions into the classifier emotions group (CEG).

6) If IEG is an empty set, assign all 17 emotions to IEG.

7) For all entries in IEG, find the shortest distances between KW1 and each emotion vector in IEG, and between KW2 and the same emotion vectors in IEG.

8) Pick the lowest distance. A list of emotion labels is shown in Table 6.2. Index I is defined as the (I + 1)th index in the array of emotion; 0 <= I <= 16. (If there is a tie, pick the emotion with the lower index).

9) Return the index of the emotion with the lowest distance.

Affect	Abbreviation	Index	Affect	Abbreviation	Index
Admiration	ADM	0	Irritation	IRR	9
Amusement	AMU	1	Joy	JOY	10
Anger	ANG	2	Pleasure	PLE	11
Anxiety	ANX	3	Pride	PRI	12
Contempt	CON	4	Relief	REL	13
Despair	DES	5	Sad	SAD	14
Disgust	DIS	6	Surprise	SUR	15
Fear	FEA	7	Tenderness	TEN	16
Interest	INT	8			

Table 6.2. *Labels for all emotions*

The output returned is the most likely emotion that the speaker seeks to convey in their utterance. These steps are described below in further detail. Figures 6.6 and 6.7 depict the flow diagram of what happens in this process. The inputs are the top X results from the classifier and the two action keywords extracted from the scenario description.

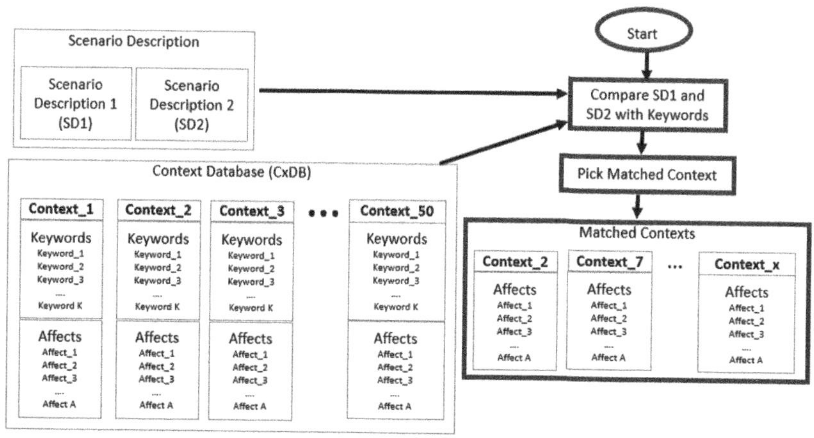

Figure 6.6. *Flowchart of CxBPARS architecture process (part 1)*

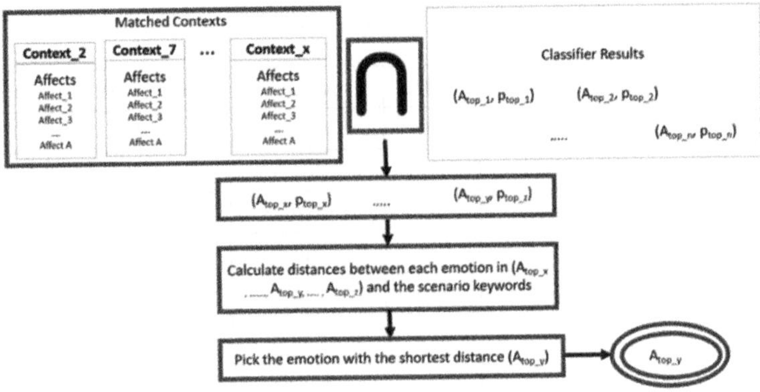

Figure 6.7. *Flowchart of CxBPARS architecture process (part 2)*

In the search to match the keywords to those in the contexts in the CxBD, we define a matched context as one that has at least one keyword in common, based on the C++ strcmp() function. By comparing the scenario keywords with the context keywords, several matched contexts can be produced. From these matched contexts, the possible emotions are extracted. These emotions are then defined as the context emotions group (CxEG). To avoid redundancy, any duplicate emotions are eliminated from the CxEG. The intersection between emotions CEG and CxEG are identified, and this intersection is defined as the intersection emotion group (IEG). To clarify, the CxEG, CEG and IEG are lists of emotion indexes (with a maximum of 17 entries), between 0 and 16. The indexes between 0 and 16 represent the full 17 emotions. These specific labels are shown in Table 6.2. Each emotion (in IEG) has its emotion vector associated with it.

This scenario assumes that there are some intersections between CEG and CxREG, and that IEG is not the null set. However, if IEG is indeed the null set, then all 17 emotions are inserted as entries in the IEG, thereby in effect nullifying the work of the phase I classifier as well as of the context database. Once IEG is produced, the distances between KW1 and all emotions in IEG are calculated, then the distances between KW2 and all emotions in IEG are likewise computed. The process then takes the lower one of the two lowest distances (one for KW1 and the other for KW2). The formula is as follows:

SemiFinal_Distance[i] = MIN (Distance (KW1, IEG[i].emotion_vector),

Distance (KW2, IEG[i].emotion_vector))

where i is defined as the index between 0 and the number of entries in IEG (between 0 and how many ever there are in IEG, with a maximum value of 16). The Distance() function measures the Euclidian distance between two points in the SS.

Subsequently, the shorter distance from the two SemiFinal_Distance[i] is identified and its index i is noted. Finally, the emotion associated with the ith entry for IEG is the output of our process that represents the emotion most likely to be felt by the speaker for a given combination of utterance and context. The output message will map the winning index to the actual name of the winning emotion to display the latter.

6.6. Example application: smart phone

In this section, we provide a detailed example of how the modeling process works. As discussed earlier, an argument could be made that the GEMEP database scenario descriptions that we used to assess our CxBPARS prototype are artificial and not directly pertinent to our daily lives. However, in reality, these scenario descriptions could roughly mimic the information that we could find in someone's smart phone calendar application, albeit in a possibly more cryptic form. For this case, we assume that these descriptions would be entered into the calendar app by a detail-oriented person who always keeps their daily routine in a very organized and highly descriptive manner in their smart phone calendar.

Much contextual information can be extracted from the smart phone calendar app entry. Some of these include a description of the event (in varying degrees of detail), the day and time of the event and the user's geolocation at the time of the event, among other things. For our current work, however, only the event description is being used. As an example, from the calendar application, we could find an entry such as: *10 a.m. – Go bungee jumping*. Next, we want to show how our CxBPARS employs both acoustic features and contextual knowledge to predict what emotion the speaker is trying to convey.

6.6.1. *Phase 1: context-free process*

Assuming a person enters an appointment/event, such as the one above, into their smart phone calendar, and later, during the event, the same phone captures the person screaming "AAHHHHH…" with high valence and high arousal. Using the acoustic features of this audio file, the classifier from phase I produces the following emotions as the top five emotions (X was set to 5): anxiety, fear, anger, amusement and despair.

6.6.2. *Phase 2: context-centered process*

Step 1 – Keyword extraction: Figure 6.8 shows the flow diagram of what happens in the keyword selection process. The input to and the output of this process, and the procedures executed in this process are as follows.

Input: an input description from a smart phone calendar or appointment app.

Procedures: 1) from this smart phone calendar app, extract the sentence from the event title: "*10 a.m. – Go bungee jumping*"; 2) from the event description, two action words (or verbs) are manually extracted. For our example, the extracted keywords are *go* and *jumping*. Therefore, the outputs of this step in the process are the keywords "go" and "jump".

Output: two scenario descriptions as action words: *go* and *jump*.

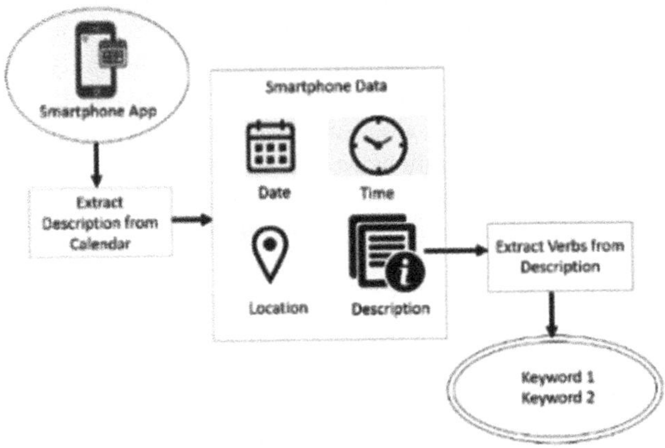

Figure 6.8. *Keyword selection process*

Step 2 – Linking keywords to context: *with these two keywords, we searched our CxDB and found a matched entry.*

Context: visit an amusement park

Keywords: go, jump, ride, experience, visit, see

Emotions: fear, joy, anxiety, amusement, pleasure

Step 3 – Word-affect mapping*:* the process from phase I produced anxiety, fear, anger and despair as predicted emotions. Step 2 produced fear, joy, anxiety, amusement and pleasure by matching the chosen keywords with the CxDB. Both processes then had these emotions in common: fear and anxiety. Next, we had to measure the distances between the keywords with these common emotions.

The distance between “go” and “fear” was 7.13 units of distance in the SS while between “jump” and “fear” was 8.10 units. And the distance between “anxiety” and “go” was 7.22 and between “anxiety” and “jump” was 8.17. Because CxBPARS takes the minimum distance between the two keywords and a particular emotion, we then took the minimum distance for each emotion. For our example, the minimum distance was 7.13 for “fear” and 7.22 for “anxiety”. In our example, “fear” would be chosen by CxBPARS as the winner – that is, the emotion most likely to be felt by the speaker.

6.7. Summary and conclusion

This chapter describes and discusses how context is used to automatically assist in identifying the emotion felt by a speaker whose words we cannot understand and whom we cannot see. We call this process context-based paralinguistic affect recognition. Our work sought to use context to improve the accuracy of the state-of-the-art paralinguistic affect recognition systems, which are all based on applying classifiers of different types to the sounds emitted by the speaker. Our work uses sentiment analysis dictionaries to relate keywords and emotions in a semantic space drawn on a two-dimensional Cartesian plane. We specifically use the dictionary by Kuperman et al. (2014). The work described here represents the second part (phase II) of the context-based paralinguistic affect recognition system. Phase I is composed of multiple classifiers on a context-free basis and is only briefly described.

We have not report any results in this chapter. The reason for this is that they are only meaningful in a full consideration of the entire system, as we did not test phase II by itself. Instead, we discuss an example of how our approach (phase II only) would be applied.

However, it is impossible to prove or contradict our hypothesis without some evidence of success or failure. Recall from section 6.2 that the hypothesis we set out was:

> Knowing the context in which that person finds themselves can be helpful in identifying their emotional state when we otherwise only have paralinguistic data.

Therefore, we feel obliged to report here that compared to the state of the art (Gosztolya et al. 2013), our CxBPARS system, which combines phases I and II, achieved a 58.4% success rate with 17 emotions, compared to Gosztolya et al.’s 42.3% success rate on the same GEMEP database. Moreover, Gosztolya et al.’s work only worked with 12 emotions – a significantly easier task than the 17 we used in CxBPARS. Gosztolya et al. only used context-free classifiers in their work. For a full description of our tests, refer to Marpaung (2019).

Given the above results, we can confidently state that our hypothesis has been validated. Nevertheless, we are the first to admit that our work cries out for further research to eliminate some of the simplifying assumptions we made in order to prove or disprove our hypothesis. These include creating the context database completely by hand – an onerous process for a database that would cover the hundreds or thousands of possible contexts. This will be our first priority in our future research. The second limitation to be lifted is how to select the keywords. The use of strcmp() is much too simple where there exist other, more effective approaches to identify similarities. Furthermore, selecting keywords other than verbs and preferably key phrases would represent an improvement.

One other major area of improvement would be to have CxBPARS (phase II) understand the implications of the situation descriptions rather than merely selecting individual keywords. That is, it would understand the meaning of phrases. Therefore, application of natural language understanding could be used to significantly improve the context modeling process. This will also be one of our priorities in our future research.

6.8. References

Banziger, T. and Scherer, K.R. (2010). Introducing the Geneva multimodal emotion portrayal (GEMEP) corpus. In *Blueprint for Affective Computing: A Sourcebook*, Scherer, K.R. Banziger, T., Roesch, E.B. (eds). Oxford University Press, Oxford.

Baroni, M. and Lenci, A. (2010). Distributional memory: A general framework for corpus-based semantics. *Computational Linguistics*, 36(4), 673–721.

Bestgen, Y. and Vincze, N. (2012). Checking and bootstrapping lexical norms by means of word similarity indexes. *Behavior Research Methods*, 44, 998–1006.

Bradley, M.M. and Lang, P.J. (1999). Affective norms for English words (ANEW): Instruction manual and affective ratings (1–45). Technical Report C-1, The Center for Research in Psychophysiology, University of Florida, FL.

Breazeal, C. (2002). *Designing Sociable Robots*. MIT Press, Cambridge, MA.

Chwelos, G. and Oatley, K. (1994). Appraisal, computational models, and Scherer's expert system. *Cognition and Emotion*, 8(3), 245–257.

Cortes, C. and Vapnik, V. (1995). Support-vector networks. *Machine Learning*, 20(3), 273–297.

Esuli, A. and Sebastiani, F. (2006). SentiWordNet: A publicly available lexical resource for opinion mining. *Proceedings of the Fifth International Conference on Language Resources and Evaluation (LREC'06)*. Genoa.

Gosztolya, G., Róbert, B., László, T. (2013). Detecting autism, emotions and social signals using adaboost. *Proceedings of the INTERSPEECH 2013 Conference*, 220–224.

Hearst, M.A. (1992). Automatic acquisition of hyponyms from large text corpora. *Proceedings of the 14th Conference on Computational Linguistics*, Nantes, 539–545.

Hollister, D., Gonzalez, A., Hollister, J. (2019). Contextual reasoning in human cognition and the implications for artificial intelligence systems. *Modeling and Using Context*, 3(1). doi: 10.21494/ISTE.OP.2019.0400.

Huang, K., Wu, C., Hong, Q., Su, M., Chen, Y. (2019). Speech emotion recognition using deep neural network considering verbal and nonverbal speech sounds. *Proceedings of the 2019 IEEE International Conference on Acoustics, Speech and Signal Processing (ICASSP)*, Brighton, 5866–5870.

Kaushik, L., Sangwan, A., Hansen, J.H.L. (2017). Automatic sentiment detection in naturalistic audio. *IEEE/ACM Transactions on Audio, Speech, and Language Processing*, 25(8), 1668–1679.

Kaza, K., Psaltis, A., Stefanidis, K., Apostolakis, K., Thermos, S., Dimitropoulos, K., Daras, P. (2016). Body motion analysis for emotion recognition in serious games [Online]. Available at: https://www.iti.gr/iti/files/document/publications/Body%20Motion%20Analysis%20for%20Emotion%20Recognition%20in%20Serious%20Games.pdf.

Kousta, S.T., Vigliocco, G., Vinson, D.P., Andrews, M., Del Campo, E. (2011). The representation of abstract words: Why emotion matters. *Journal of Experimental Psychology: General*, 140(1), 14–34.

Kuperman, V., Estes, Z., Brysbaert, M., Warriner, A.B. (2014). Emotion and language: Valence and arousal affect word recognition. *Journal of Experimental Psychology: General*, 143, 1065–1081.

Leveau, N., Jhean-Larose, S., Denhière, G., Nguyen, B. (2012). Validating an interlingual metanorm for emotional analysis of texts. *Behavior Research Methods*, 44, 1007–1014.

Liu, B. (2012). Sentiment analysis and opinion mining. *Synthesis Lectures on Human Language Technologies*, 5(1), 1–167.

Marpaung, A. (2019). Context-centric affect recognition from paralinguistics features of speech. PhD Dissertation, University of Central Florida, Orlando, FL.

Marpaung, A. and Gonzalez, A. (2014). Toward building automatic affect recognition machine using acoustics features. *Proceedings of the 27th International FLAIRS Conference*, Pensacola Beach, FL.

Marpaung, A. and Gonzalez, A. (2017). Can an affect-sensitive system afford to be context independent? *Proceedings of the International and Interdisciplinary Conference on Modeling and Using Context (CONTEXT 2017)*, 454–467.

Marpaung, A. and Gonzalez, A. (2021). Role of contextual knowledge in paralinguistic affect recognition [Online]. Available at: https://www.openscience.fr/Role-of-Contextual-Knowledge-in-Paralinguistic-Affect-Recognition-by-Humans.

Meng, H., Yan, T., Yuan, F., Wei, H. (2019). Speech emotion recognition from 3D log-Mel spectrograms with deep learning network. *IEEE Access*, 7, 125868–125881.

Miller, G.A. (1995). WordNet: A lexical database for English. *Communications of the ACM*, 38(11), 39–41.

Mohammad, S.M., Zhu, X., Kiritchenko, S., Martin, J. (2015). Sentiment, emotion, purpose, and style in electoral tweets. *Information Processing & Management*, 51(4), 480–499.

Noroozi, F., Kaminska, D., Corneanu, C., Sapinski, T., Escalera, S., Anbarjafari, G. (2018). Survey on emotional body gesture recognition. *IEEE Transactions on Affective Computing*, doi: 10.1109/TAFFC.2018.2874986.

Opitz, D. and Maclin, R. (1999). Popular ensemble methods: An empirical study. *Journal of Artificial Intelligence Research*, 11, 169–198.

Pereira, C., Zaslavsky, A., Christen, P., Georgakopoulos, D. (2014). Context aware computing for the internet of things: A survey. *Communications Surveys Tutorials, IEEE*, 16(1), 414–454.

Pérez-Rosas, V., Kleinberg, B., Lefevre, A., Mihalcea, R. (2017). Automatic detection of fake news. arXiv preprint arXiv:1708.07104.

Picard, R.W. (1997). *Affective Computing*. MIT Press, Cambridge, MA.

Sailunaz, K., Dhaliwal, M., Rokne, J., Alhajj, R. (2018). Emotion detection from text and speech: A survey. *Social Network Analysis and Mining*, 8(1), 1–26.

Schapire, R. and Freund, Y. (2012). *Boosting: Foundations and Algorithms*. MIT Press, Cambridge, MA.

Scherer, K.R. (1993). Studying the emotion-antecedent appraisal process: An expert system approach. *Cognition and Emotion*, 7, 325–355.

Scherer, K.R. (1999). Appraisal theory. In *Handbook of Cognition and Emotion*, Dalgleish, T. and Power, M. (eds). John Wiley and Sons, Chichester.

Schmidhuber, J. (2015). Deep learning in neural networks: An overview. *Neural Networks*, 61, 85–117.

Schuller, B., Zhang, Y., Weninger, F. (2018). Three recent trends in paralinguistics on the way to omniscient machine intelligence. *Journal on Multimodal User Interface*, 12(4), 1–11.

Soleymani, M., Garcia, D., Jou, B., Schuller, B., Chang, S.-F., Pantic, M. (2017). A survey of multimodal sentiment analysis. *Image and Vision Computing*, 65, 3–14.

Stuhlsatz, A., Meyer, C., Eyben, F., Zielke, T., Meier, G., Schuller, B. (2011). Deep neural networks for acoustic emotion recognition: Raising the benchmarks. *Proceedings of ICASSP*, IEEE, Prague, 5688–5691.

Trigeorgis, G., Ringeval, F., Brückner, R., Marchi, E., Nicolaou, M., Schuller, B., Zafeiriou, S. (2016). Adieu features? End-to-end speech emotion recognition using a deep convolutional recurrent network. *Proceedings of ICASSP*. IEEE, Shanghai, China, 5200–5204.

Verona, E., Sprague, J., Sadeh, N. (2012). Inhibitory control and negative emotional processing in psychopathy and antisocial personality disorder. *Journal of Abnormal Psychology*, 121(2), 498–510.

Wehrle, T. and Scherer, K.R. (1995). Potential pitfalls in computational modelling of appraisal processes: A reply to Chwelos and Oatley. *Cognition and Emotion*, 9(6), 599–616.

Wöllmer, M., Eyben, F., Reiter, S., Schuller, B., Cox, C., Douglas-Cowie, E., Cowie, R. (2008). Abandoning emotion classes – Towards continuous emotion recognition with modelling of long-range dependencies. *Proceedings of INTERSPEECH*, ISCA, Brisbane, 597–600.

Yadollahi, A., Gholipour Shahraki, A., Zaiane, O. (2017). Current state of text sentiment analysis from opinion to emotion mining. *ACM Computing Surveys CSUR*, 50(2), 1–33.

Yu, Z., Liu, G., Liu, Q., Deng, J. (2018). Spatio-temporal convolutional features with nested LSTM for facial expression recognition. *Neurocomputing*, 317, 50–57.

Zeng, N., Zhang, H., Song, B., Liu, W., Li, Y., Dobaie, A. (2018). Facial expression recognition via learning deep sparse autoencoders. *Neurocomputing*, 27, 643–649.

Zhang, S., Zhang, S., Huang, T., Gao, W. (2016). Multimodal deep convolutional neural network for audio-visual emotion recognition. *Proceedings 6th ACM International Conference Multimedia Retrieval (ICMR)*, New York, 281–284.

Zhang, S., Zhang, S., Huang, T., Gao, W., Tian, Q. (2018). Learning affective features with a hybrid deep model for audio–visual emotion recognition. *IEEE Transactions on Circuits and Systems for Video Technology*, 28(10), 3030–3043.

Zhao, J., Mao, X., Chen, L. (2019). Speech emotion recognition using deep 1D & 2D CNN LSTM networks. *Biomedical Signal Processing and Control*, 47, 312–323.

7

Context-Driven Behavior: A Proactive Approach to Contextual Reasoning

Over the last three decades, a considerable amount of research has been dedicated to improving an artificial agent's ability to recognize and effectively deal with context. In this chapter, we discuss a framework for a novel form of contextual reasoning. Unlike existing contextual reasoning frameworks, which allow an agent to apply its contextual knowledge after it is operating in an instance of a known context, the model we discuss allows an agent to reason about context proactively. With a proactive model, an agent makes predictions about the future contexts it will encounter and then takes steps to ensure that its behaviors are appropriate for those contexts before entering into them.

7.1. Motivation for a proactive model

There can be no serious objection to the statement that for humans and artificial agents alike, behavior is context-dependent. For example, a person who hopes to become gainfully employed would not consider text messaging a friend to discuss weekend plans while attending a job interview. Why not? Because most people know (either through social inheritance or from prior experience) that this type of behavior is not appropriate while they are in the context "attending an interview".

For an artificial agent, context-appropriate behavior is just as important as it is for humans. After all, an agent that is unable to behave appropriately for the contexts it will operate in is essentially useless. Over the last three decades, there has been a considerable amount of research dedicated to the development of models and frameworks for improving an artificial agent's ability to recognize and effectively

---

Chapter written by Christian WILSON.

deal with context. What many of these models have in common is they are primarily "reactive" in nature. By reactive, we mean an agent will modify or adapt its behaviors to a specific context only after it is operating in a known instance of that context (Gonzalez and Ahlers 1998; Turner 1998).

The ability to react to contextual changes is a critical component of any contextual reasoning framework because an agent's situation can change unexpectedly. When these changes occur, the agent must be able to identify any new contexts and adjust its behaviors accordingly. Consider, for example, an autonomous underwater vehicle that is on a mission in the open ocean. While on the mission, the AUV could lose propulsion or its sonar could fail. When these types of events occur, the AUV should determine which context it is operating in and how it should behave. For example, when in the context "no propulsion", the only appropriate behavior might be to surface and wait for a rescue vessel to come to its aid.

One drawback to a purely reactive model, however, is no consideration for future contexts is given. This means that there is a brief period of time when a context is active and context-appropriate behavior is expected, but the agent is reasoning about how it should behave. For many agents, a purely reactive model may be sufficient. An agent might operate in an environment where a delayed response to contextual changes will have no effect on the agent's performance or its observable behaviors.

In some cases, a reactive model could result in context-inappropriate behavior. Depending on the application or environment, that behavior could have catastrophic consequences. Suppose, for example, an AUV is on a mission to map a live minefield. When in the context "in a minefield", the AUV should reduce its speed and turn off its sonar so as to avoid colliding with or triggering a mine. Ideally, the AUV would make these types of context-specific adjustments to its behavior before entering into the minefield. With a reactive model, however, there is an increased risk of accidental detonation because the AUV will not make any adjustments to its behavior until after it has entered the minefield. While a catastrophe could be avoided by adding appropriate actions to the AUV's mission plan, this solution is contrary to what most contextual reasoning frameworks aim to achieve.

As the previous example demonstrates, a model of contextual reasoning should be robust and allow an agent to reason about context in different ways. When contextual changes occur unexpectedly, the model should allow an agent to react appropriately to those changes. The model should also allow the agent to look ahead and reason about future contexts, especially in situations where contextual changes are expected. When on a mission to map a minefield, for example, the agent should know that at some point during the mission, it will be operating in the context "in minefield". With an awareness of this future context, the agent should use its knowledge of this context to augment its mission plan with actions (e.g. reducing its speed, turning off sonar, etc.) that allow it to enter this context cleanly.

In this chapter, I will describe a framework for a proactive model of contextual reasoning. I will refer to this model as *context-driven behavior* (CDBr). Unlike a reactive model in which an agent uses its contextual knowledge to decide how to behave while in a context, CDBr allows an agent to reason about future contexts, then use its knowledge of those contexts to "drive" aspects of its behavior. A formal definition of context-driven behavior is given below.

DEFINITION 7.1.– Context-driven behavior (CDBr) is the process of reasoning about the context transitions and future contexts that will occur during a problem-solving episode, then adapting a problem-solving strategy to those contexts. This can include modifying certain behaviors before a particular context activates, modifying components of problem-solving strategy to ensure behaviors are appropriate for those contexts and in some cases, modifying instances of contexts themselves to allow for context-appropriate goal accomplishment. A context-driven agent is an agent that is capable of exhibiting this type of behavior.

The remainder of this chapter is organized as follows. First, I will address some of the challenges associated with a proactive model of contextual reasoning. Next, I will talk about context and contextual knowledge as it pertains to context-driven behavior. Finally, I will discuss the components of an initial framework for achieving context-driven behavior in an artificial agent and how that framework attempts to address the challenges associated with a proactive model of contextual reasoning.

7.2. Challenges associated with a proactive model

With a reactive model of contextual reasoning, an agent reasons about context based on what it knows about the features of its current situation. The agent observes situational changes then determines its current context based on what it has observed. With a proactive model, an agent is instead basing its reasoning on situational changes that *could* or *might* occur at some point in the future. This means that the agent must make accurate predictions about future situations if it is going to make accurate predictions about the contexts it will encounter.

7.2.1. *Coping with uncertainty*

One of the biggest challenges associated with a proactive model of contextual reasoning is coping with uncertainty. For example, in closed or "toy" domains such as those used to test the performance of an automated planner, an agent will typically have a complete understanding of its world. The agent can observe all features of its situation and knows how its actions will affect those features. Additionally, in most toy domains, there are no other agents, processes or events that can perturb the agent's world in a manner that is unexpected. Therefore, in closed domains, an agent can predict the state of its future situation and, ultimately, any future contexts, with near-perfect confidence.

In "open" domains such as the real world, and similarly, virtual domains intended to simulate the real world (e.g. video games, social and military simulations, etc.), predicting future situations becomes more difficult for several reasons, all of which are interrelated to some degree. First, in open domains, an agent will seldom have complete information about its world, the goals it is pursuing or even itself. This is due to the fact that no agent will have the resources needed to observe its world completely, even at a local scale. It is also unlikely that a human will have enough knowledge about the agent's world to give to the agent. Even if an agent were capable of observing its world completely, organizing, storing and updating that knowledge so it can be accessed when needed would likely consume the agent's limited computational resources.

Second, in open domains, an agent must cope with the presence and actions of other agents that inhabit its world. In general, an agent will not know about the goals and intentions of other agents or the effects their actions will have on its world. For example, after hearing the news of a novel virus outbreak, a person might make plans to go to a local grocery to purchase basic goods in the event they are quarantined due to the outbreak. What the person may not know is that supplies of basic goods such as canned goods, flour and even toilet paper have been depleted because the news of the pandemic has resulted in other people hoarding those goods.

Finally, open domains are often unpredictable. Events such as accidents, mechanical malfunctions or acts of nature can occur unexpectedly. For example, while on a mission in the open ocean, it is possible for an AUV to experience a sudden mechanical or electrical failure or become entangled in a fishing net that detached from a nearby fishing vessel. It can be difficult if not impossible for an agent to predict contexts related to these types of events because their occurrence is outside the scope of its control.

All agents will face some form of uncertainty as they work to achieve their goals. With context-driven behavior, a great deal of uncertainty stems from the fact that the agent is reasoning about the contexts it will encounter at some time in the future. In general, an agent can find itself in any number of active contexts at a given moment. For example, on a given day, a person might find themselves in the contexts "at work", "inclement weather", "in a meeting" and "stock market crash". For most people, contexts such as “at work” or “in a meeting” might be expected because they are part of a person’s daily routine. If the person is not an avid investor or otherwise possesses some form of specialized knowledge, it is unlikely that they would be able to predict being in a context such as “stock market crash”.

A context-driven agent does not need to predict all future contexts but rather only those contexts that will have immediate implications for its behavior. As a person prepares to commute to work, for example, their behaviors are likely influenced by future contexts such as “at work”. That is, we expect the person to choose and dress in clothes that are appropriate for this context and perhaps perform other actions that ensure they will enter this context cleanly and function successfully while in that context. For example, if the person knows that they will have a limited amount of time for a lunch and food is not easily accessible while in the context “at work”, they might also pack some food to eat or make plans to purchase food during the commute. When commuting to work, it is possible that the person might find themselves in the context “in traffic jam” or worse, “in automobile accident”. However, it is not reasonable to expect the person’s behaviors to be influenced by these contexts regardless of how likely they are to occur.

To reduce the amount of uncertainty a context-driven agent must cope with during context prediction, the CDBr framework restricts context prediction to the contexts associated with the goal or goals the agent is pursuing. Exogenous and contemporaneous contexts will be ignored. Restricting the scope of context prediction has several benefits. First, exogenous or contemporaneous contexts may have little to no effect on agent behaviors. For example, if a person is not an avid investor or planning for retirement, the context “stock market crash” may have little to no impact on the person’s behavior. Omitting these types of contexts from context prediction is intended to conserve an agent’s computational resources. Second, exogenous and contemporaneous contexts are often related to situational changes outside the scope of an agent’s control. Predicting situational changes related to the actions of other agents may not be computationally feasible depending on the type of environment the agent is operating in.

Restricting the scope of context prediction to goal-related contexts reduces the problem of situation prediction in open domains to a problem in which an agent can make reasonable predictions features of a future situation using knowledge of the effects its actions will have on its environment. It is worth noting that ignoring future contexts besides those related to the goal or goals the agent is pursuing does not hinder a context-driven agent’s ability to effectively deal with those contexts if and when they are encountered. An agent should have more than one method for dealing with contextual changes if and when they occur, and these types of contexts are ideal candidates for a reactive model of contextual reasoning.

7.2.2. *A lack of initial knowledge*

Humans are an ideal example of a context-driven agent. Our behavior would suggest that we take future contexts into consideration as we work to achieve our goals and we use knowledge of those contexts to guide aspects of our behavior. For example, we would not expect a person to commute to work, and then once they are at work, determine what type of clothing is appropriate for this context. Instead, we expect the person to dress in context-appropriate clothing before commuting to work or, alternatively, before arriving at work.

A person's apparent awareness of the future contexts that they will encounter is likely based on a wealth of knowledge that has been acquired and refined over time. For example, a person might associate contexts such as "in public" or "not at home" with any goal that is pursued outside the home. Additionally, a person might associate more specific contexts such as "at work", "in a meeting" or "attending class" with other aspects of daily living. Unfortunately, an artificial agent will not have the same type or amount of contextual knowledge that most people have.

While it is possible for a knowledge engineer to provide an agent with some form of initial knowledge that allows it to anticipate future contexts, manually encoding this type of knowledge is not a tractable solution. There can be any number of future contexts an agent might encounter. Some of those contexts might be related to the agent's environment. Others might be related to the agent's own actions or the actions of other agents operating in the same environment. Maintaining an accurate reference to possible contexts and the conditions under which the agent will encounter those contexts will quickly become unmanageable.

Restricting context prediction to goal-related contexts does little to reduce the burden associated with representing or encoding knowledge of future contexts in an artificial agent. As I discuss in the following sections, there can be a variety of contexts associated with a goal the agent is pursuing. Some of those contexts are related to the goal itself, whereas others are related to the agent's strategy for achieving that goal. Defining an association between goals and possible contexts can become equally unmanageable, especially for agents that have multiple goals and, additionally, multiple strategies for achieving each goal.

To reduce the burden placed on a knowledge engineer and to reduce the need for highly specialized contextual knowledge, the CDBr framework allows an agent to make predictions and reason about future contexts using a reasonable amount of contextual knowledge that would otherwise allow an agent to reason about those same contexts in a reactive manner.

7.3. Context and contextual knowledge

Context is a term that can have numerous definitions. Even within a single domain, authors often provide a definition of context that is tailored to their particular research. Rather than introducing yet another definition of context into an overly crowded field, I will instead adopt the definition of Turner (1998) in his work on context-mediated behavior (CMB). Turner defines context as "any identifiable configuration of environmental, goal-related and agent-related features that has predictive power for an agent's behavior".

The decision to adopt Turner's definition is not simply for the sake of convenience. First, defining a context in terms of situational features alone means that a knowledge engineer can define contexts for the situations an agent will operate in. Second, this definition allows multiple contexts to be defined for a given situation. This is critical because an agent will often find itself operating in more than one context at any given time. Finally, defining a context in terms of situational features means that contextual changes can be defined in terms of changes to those features. The latter is critical because a context-driven agent must be able to reason about contextual changes.

7.3.1. *Problem-solving contexts*

When an agent commits to a goal and executes a plan to achieve that goal, we say that the agent is engaged in a *problem-solving episode*. During a problem-solving episode, the agent will transition through a variety of active contexts related to the current problem-solving episode. We will refer to these contexts as *problem-solving contexts* (PSCs).

CDBr makes a further distinction between two types of PSCs, namely, *goal-related* and *strategy-related* PSCs. Goal-related PSCs are problem-solving contexts, which are directly related to the goal or goals the agent is pursuing. When satisfying hunger, for example, a person will at some point find themselves in the context "eating food" or "eating a meal" regardless of whether they are eating a meal at home or dining at a restaurant with friends.

Strategy-related PSCs are problem-solving contexts that are a consequence of the agent's strategy for achieving a goal. As such, the strategy-related PSCs that an agent encounters can differ depending on the chosen strategy. If a person is eating a meal at home, they might transition through the strategy-related PSCs "preparing food" and "cooking on the stove" before entering the context "eating a meal". If the person is dining out with friends, they might transition through the strategy-related

PSCs "driving a car", "in public" and "at restaurant" before entering the context "eating a meal".

Making a distinction between different types of active contexts is crucial because each type of context can have different implications for an agent's behaviors. For example, because goal-related PSCs are related to the goal an agent is pursuing, there is little the agent can do to avoid these contexts if it chooses to commit to a goal. As such, recognizing goal-related contexts will allow an agent to assess the appropriateness of a candidate goal. However, if the agent wants to avoid a strategy-related context, the agent can determine if there is an alternative strategy for achieving a goal.

7.3.2. *Contextual schemas*

All agents have some form of contextual knowledge, whether implicit or explicit. With a context-driven agent, contextual knowledge is represented explicitly. We will not talk in any detail about the benefits of explicitly representing contextual knowledge since that topic is adequately covered elsewhere (Turner 1994).

The structure used to represent an agent's contextual knowledge is an extended version of the contextual schema (c-schema), originally proposed by Turner (1999). C-schemas are declarative knowledge structures that contain both descriptive and prescriptive knowledge about the context they represent. Descriptive knowledge includes features of the situation that must be present (or not present) in order for it to be considered an instance of the context. This information allows the agent to "diagnose" the current situation and determine if it is an instance of a context it knows about. For example, descriptive knowledge for the context "at restaurant" might tell an agent that it will be in a public setting, typically a building. The building will have a kitchen, bathrooms and large dining rooms with tables and chairs where food is served.

Features of the situation, perhaps yet unseen, that are expected in this context can also be included in this knowledge. These features allow the agent to make predictions about things it is likely to see that may impact problem-solving (e.g. to allow it to recognize unanticipated events). Continuing with the previous example, descriptive knowledge should tell an agent that there are a finite number of tables and chairs available to patrons. If all tables are occupied, the agent will have to wait for an available table before it can be seated.

Descriptive knowledge can also include context-specific ontology/meaning of certain concepts because some concepts have different meanings in different contexts. For example, the concept "too deep" can have different meanings depending on whether an AUV is operating in a harbor or on a mission in the open ocean.

Prescriptive knowledge tells an agent how to behave while the context is active. Prescriptive knowledge can tell an agent which goals are important, which goals should be pursued and which goals should be avoided. For example, prescriptive knowledge for the context "at restaurant" might tell an agent that goals such as balance checkbook or schedule doctor's appointment are not important while this context is active and goals such as make personal phone call should be avoided because they are not appropriate for this context.

Prescriptive knowledge can also tell an agent how to achieve goals in a manner that is appropriate for the current context. For example, how an agent achieves the goal satisfy hunger, depends on its current situation. If the agent is at home, it can navigate to the kitchen and help itself to any food that if finds available. While in the context "at restaurant", however, it must first order food from a waiter or waitress and then wait for the food to be prepared and served.

To demonstrate the type of information represented in a traditional c-schema, an example of a partial c-schema for the context "at restaurant" is shown in Figure 7.1.

```
Actors:
    *waiter isa intelligent agent ;; one or more waiters
    *waitress isa intelligent agent ;; one or more waitresses
    *patron isa intelligent agent ;; zero or more patrons
Setting:
    day: any
    time: any
    place:
      isa: locale
      descriptor: a public building
      environment:
        dining room: isa physical space
        bathroom: isa physical space
        kitchen: isa physical space
        objects:
          table: isa physical object ;; zero or more tables for dining
          chair: isa physical object ;; zero or more chairs for dining
Events:
    introduction to waiter or waitress, seating, meal selection, ordering,
    food preparation, food served, dining, ...
Goals:
    satisfy hunger
...
```

Figure 7.1. *Partial c-schema for the context "at restaurant"*

7.3.2.1. *Problematic features*

An active context will elicit certain behaviors from an agent. Sometimes, desired or expected behaviors are inherent to the context itself as a result of a societal or cultural norm. For example, no sane person would walk down a public or city street without clothing because this behavior is not accepted in most societies. Therefore, most people know that appropriate attire is required while in the context "in public" or "not at home".

Sometimes, the features that define a context will, themselves, have implications for an agent's behaviors. For example, while in the context "cooking on the stove", a person will likely want to avoid burning the food or worse, starting a kitchen fire. As a result, the person might remain at the stove or in the kitchen and avoid any goal or activity that takes their attention away from the cooking task. In this case, the person's behaviors should be motivated by the fact that the stove is on and there is food cooking on the stove.

In this work, a situational feature that has implications for an agent's behavior will be referred to as a *problematic feature*. When a context is composed of problematic features, information about those features will be represented in the corresponding c-schema as a type of descriptive knowledge. Any feature constraints related to a problematic feature will also be included in the c-schema as a form of prescriptive knowledge. As shown in Figure 7.2, this information describes each problematic feature and includes an additional reference to actions that can potentially remove those features from the agent's situation. Including a reference to actions that remove a problematic feature from the situation provides an agent with context-specific instructions for modifying the context should the need arise.

7.3.2.2. *Feature constraints*

A traditional c-schema can include information about the goals that are considered appropriate or alternatively, inappropriate, while operating in an instance of the context. For example, a c-schema for the context "cooking on the stove" might inform an agent that goals such as taking a nap or watching television are considered inappropriate while this context is active.

Hard-coding information about appropriate or inappropriate goals in an agent's c-schemas provides a quick and convenient method for representing that information, but it is not without limitations. First, for some contexts, there can be any number of goals that may be considered inappropriate while that context is active. In these cases, it may not be feasible to encode information about all inappropriate goals in a c-schema. The same limitation applies if, alternatively, information about appropriate goals were to be encoded.

```
Actors:
    self
    *others ;;zero or more intelligent agents

Setting:
    day: any
    time: any
    place:
        house: isa locale
        descriptor: a private dwelling
    location:
        kitchen: isa physical space
        descriptor: for storing, preparing, cooking and eating food
    objects:
        stove: isa physical object
        burner: isa physical object ;; a heating element on the stove
        pot: a physical object ;; one or more pots for cooking
        pan: a physical object ;; one or more pans for cooking
        food: a physical/edible object ;; one or more food items
        ...

Problematic features:
    feature: burner is on
        descriptor: state of the burner
        clean-up: turn off burner
    feature: pan is on burner
        descriptor: state of the pan
        clean-up: remove pan from hot burner
    feature: pan contains food
        descriptor: state of the food items
        clean-up: remove pan from hot burner

Feature constraints:
    feature: agent is in house
        descriptor: location of the agent
        constrained-by: burner is on; pan with food on stove
        preference: 1
    feature: agent is in kitchen
        constrained-by: burner is on; pan with food on stove
        preference: 1
    feature: agent is at stove
        constrained-by: burner is on; pan with food on stove
        preference: 0
```

Figure 7.2. *Partial extended c-schema for context "cooking on the stove"*

Second, classifying a goal as inappropriate for a particular context only tells an agent to prune that goal from the space of potential goals whenever it is encountered in that context. This type of decision-making is rigid and if care is not taken it can lead to predictable patterns of observable behavior because the agent will always defer an inappropriate goal whenever it is encountered in that context. For some agents, this type of decision-making is sufficient. For others, a more appropriate course of action might be to consider an alternative plan for achieving a goal in a manner that is appropriate for those contexts.

Context-driven behavior uses an alternative method for representing information about context-inappropriate goals in an agent's c-schemas. As opposed to encoding explicit references to context-inappropriate or context-appropriate goals, information about problematic features is used to establish implicit constraints on the goals an agent can pursue whenever it is operating in an instance of that context. These constraints are referred to as *feature constraints*.

A feature constraint defines an *existential* relationship between a problematic feature and one or more non-problematic features of an agent's situation. When a feature constraint is present in a c-schema, it informs an agent that the non-problematic (i.e. constrained) features should exist in the agent's situation whenever the problematic (i.e. constraining) feature exists.

Feature constraints are designed to restrict the goals an agent pursues by restricting the situational changes that are allowed to occur whenever a problematic feature exists. For example, let us once again consider the context "cooking on the stove". As shown in Figure 7.2, this context consists of the problematic features (on stove pan) and (is-on stove). To ensure an agent remains focused on the cooking task, the c-schema establishes feature constraints between each of the problematic features and the spatial features (in-house agent), (at-kitchen agent) and (at-stove agent).

For some situations however, there are exceptions to the behaviors an agent is expected to exhibit. For example, while cooking on the stove, it is reasonable for a real human to pursue other goals in the kitchen such as washing dishes or cleaning any residual mess made during meal preparation. To allow for variability in an agent's behavior, feature constraints are assigned a numeric *preference value* $p \in \{0,1\}$.

Preference values indicate whether a constraint should be strictly adhered to (i.e. $p = 1$) or if there are situations where the constraint can be ignored (i.e. $p = 0$). Referring to Figure 7.2, the constraints on the spatial features (in-house agent) and (at-kitchen agent) are assigned a preference value of 1, whereas the constraint on the spatial feature (at-stove agent) is assigned a preference value of 0. This restricts the goals the agent pursues to those that can be achieved in the kitchen.

7.4. A framework for context-driven agent

The context-driven behavior process can be summarized as follows. When presented with one or more goals, an agent will select a candidate goal and then select a tentative strategy for achieving that goal. The agent will then attempt to predict the future contexts it expects to encounter during the problem-solving episode and determine if the candidate goal can be achieved in a manner that is appropriate for those contexts. If inappropriate behavior is detected, the agent will determine an appropriate course of action for remedying that behavior.

The framework presented in this section distributes the process outlined above across four reasoning tasks, which include defining a problem-solving scenario, context prediction, identifying context-inappropriate behavior and, finally, strategy modification. I will talk about each task in more detail below. Since CDBr is being developed to improve the quality and plausibility of virtual human behavior (Wilson and Turner 2017), most of the examples are in that domain. Where appropriate, examples from other domains will be included.

7.4.1. *Defining a problem-solving scenario*

Within the CDBr framework, a problem-solving scenario is an abstract knowledge structure that is used to store the results of the reasoning tasks throughout the CDBr process. This can include a reference to the contexts the agent expects to encounter, any problematic behavior the agent anticipates and, in some cases, steps to address that behavior. The first task in the CDBr process is defining a problem-solving scenario.

When the agent defines a problem-solving scenario, it will select a candidate goal from the space of pending goals then select a tentative plan for achieving that goal. A reference to the candidate goal and the tentative strategy is then added to an instance of a problem-solving scenario. This task is not unique to CDBr. For example, all agents have a means of prioritizing and identifying important goals then selecting a plan for achieving those goals. For more sophisticated agents, the latter might involve synthesizing a plan that achieves a candidate goal, whereas a less sophisticated agent might instead select a pre-scripted plan from a plan library. I will assume that regardless of the agent implementation, a context-driven agent has the ability to select a candidate goal and identify an initial plan for achieving that goal.

Examples in the following sections use what I will refer to as a default plan, which provides an agent with a general solution for achieving a goal. The steps in a default plan contain an action, details about preconditions that must be satisfied before the action can be successfully executed and details about postconditions the agent can assume to be true if the action is executed successfully. However, a default plan lacks context-specific information such as context-specific preconditions or actions. As a result, a default plan might allow an agent to make progress toward achieving a goal but the appropriateness of the agent's behaviors is not guaranteed. An example of a default plan for checking the mail is shown in Figure 7.3.

```
Goal: Check mail

;; Agent is at the mailbox and possesses mail
Goal-state: (and (at (mailbox agent))
                 (have (mail agent))

;; Navigate to the front/main door
step-1: (:action GOTO :parameters (door agent)
         :preconditions (in-house agent)
         :postconditions (at-door agent)))
;; Open the door
step-2: (:action OPEN :parameters (front-door)
         :preconditions (at-door agent))
         :postconditions (is-open door))
;; Exit the house
step-3: (:action EXIT :parameters (house agent)
         :preconditions (in-house agent)
         :postconditions (not (in-house agent))
;; Close the door
step-4: (:action CLOSE :parameters (door agent)
         :preconditions (is-open door)
         :postconditions (not (is-open door))
;; Navigate to the mailbox
step-5: (:action GOTO :parameters (mailbox agent)
         :preconditions (not (in-house agent))
         :postconditions (at (mailbox agent))
;; Open the mailbox
step-6: (:action OPEN :parameters (mailbox agent)
         :preconditions (at (mailbox agent))
         :postconditions (is-open mailbox)
;; Get/obtain mail
step-7: (:action GET :parameters (mail agent)
         :preconditions (and
                         (at (mailbox agent))
                         (is-open mailbox)
                         (contains (mail mailbox)))
         :postconditions have(mail agent)
;; Close the mailbox
step-8: (:action CLOSE :parameters (mailbox agent)
         :preconditions (and (at (mailbox agent) (is-open mailbox))
         :postconditions (not (is-open mailbox))
```

Figure 7.3. *Default plan: check mail*

7.4.2. *Predicting future contexts*

Before an agent can modify or adapt its behaviors to any future contexts, it must have an idea about the contexts it expects to encounter. Context prediction is a two-step process in which the agent predicts situational changes that will occur during a problem-solving episode and then identifies the contexts that will activate (or deactivate) as a result of those changes.

7.4.2.1. *Language of the situation calculus*

When talking about situations and situational changes, I will use language and notation consistent with the situation-as-histories (Levesque et al. 1998) variant of the situational calculus (McCarthy 1963). The situational calculus is a broad topic that attempts to axiomatize situations and their properties. Here, I will present the reader with a discussion of the concepts I will use throughout the remainder of this chapter. This includes situations, actions and objects.

The situation. There are two symbols for the situation, namely, the constant symbol S_0 and the binary function symbol $do(A, S) = S'$. S_0 is used to denote the initial situation, and the function $do(A, S) = S'$ represents the successor situation resulting from performing the action A in situation S. The binary predicate symbol $\sqsubset$ is used to define an ordering relation on situations. That is, $S \sqsubset S_0$ means S_0 can be reached from S with a sequence of actions. More formally, S is a proper sub-history of S_0.

The action. An action A is an n-ary function $A(f_1, f_2 \ldots f_n)$ which is applied to the features f_i of a situation S. Actions are subject to the binary predicate $Poss(A,S)$ meaning it is *possible* to perform the action A in situation S.

Properties of situations (i.e. what I refer to as features) are represented using relational fluents. In the situation calculus, a relational fluent is typically a binary relation of the form $R(O,S)$, where O is an object in situation S. For example, the fluent *IsOn*(*television*,*S*) would be used to denote the fact that the object *television* is powered on in situation S. While this notation sufficiently captures the state of certain objects in a situation, it fails to describe other properties of a situation that do not have a binary relational representation. For example, in a social simulation, an agent might be wearing various articles of clothing in a particular situation. These properties are easier to describe with a ternary fluent such as *Wearing*(*shirt*,*agent*,*S*) and *Wearing*(*pants*,*agent*,*S*). To this end, I will treat relational fluents as *n-ary* operators of the form $R(O_1 \ldots O_{n-1}, S)$, where each O_i is an object in situation S.

7.4.2.2. *Using plan projection to predict future situations*

As an agent works to achieve its goals, its actions will cause features of its situation to change. Knowledge about these changes is implicitly declared in the agent's strategy for achieving a goal. Knowledge about strategy-specific changes is encoded as *postconditions* or *effects* that describe the effects an action will have on the agent's situation after it has been executed. Referring to the default plan shown in Figure 7.3, the postconditions associated with the actions in steps 1 and 3 describe changes to the agent location associated with each action.

Unfortunately, postconditions do not provide an agent with enough information to make reasonable predictions about future situations. For example, postconditions only describe features that are modified or introduced by the actions in a given strategy. They do not include a reference to any feature that is unaffected by those actions. If the agent is going to make reasonable predictions about the state of a future situation, it must also maintain a reference to features that are unaffected by the agent's actions.

Perhaps what is more important, the steps in a problem-solving strategy are often performed according to a particular order. To maintain an accurate representation of the state of situational features over the duration of a problem-solving episode, the agent must reference postconditions in the same order. For example, the ordering of the steps shown in Figure 7.3 indicates that an agent must open the front door before exiting the house and, subsequently, the agent must exit the house and close the front door before it can proceed to the mailbox. If the agent is going to maintain an accurate reference to its location and the state of the door, the postconditions must be referenced in the same order.

Maintaining an accurate reference to the state of situational features when referencing postconditions is complicated by the fact that multiple actions can have an effect on the same situational feature. For example, in Figure 7.3, the actions in steps 2 and 4 operate on the front door; however, each action will have a different effect on the state of the door after it has been executed. This example demonstrates how the order in which situational changes are assumed to occur will have an impact on the predicted state of a future situation.

To ensure that an agent is able to make accurate predictions about future operating contexts, it is not enough to reference the postconditions in a problem-solving strategy.

Rather, a context-driven agent must be able to reference postconditions according to a particular ordering of steps and then use that information to predict the cumulative effects of the problem-solving strategy in its entirety. To this end, the CDBr framework uses a form of plan projection as a model for predicting situational changes associated with a particular plan.

Plan projection (Hanks 1991; McDermott 1997) is the process of predicting the outcome of a plan based on what is known about a situation at execution time. More specifically, plan projection is a form of temporal reasoning that allows an agent to look ahead and consider the consequences of a potential plan. The process of *projecting* as it is known in AI was originally introduced to improve the performance of automated planners with respect to various planning operations. For example, plan projection provides an agent with a mechanism for detecting problems with a potential plan or comparing the results with those of an alternative plan in an effort to determine an optimal solution. Plan projection also allows an agent to reason about the state of various facts during plan execution. For an automated planner, this type of reasoning is critical because certain planning operations such as plan reordering can be contingent on the state of those facts.

Similar to other models of plan projection, the model I use in this work begins with an initial situation S_0, which represents the agent's beliefs about the features of its situation before plan projection begins. Given an initial situation and a plan, the model will produce a temporally ordered sequence of situation histories, which describe the evolution of the initial situation under the actions of the problem-solving strategy. Each *situation history* is a partial representation of the agent's situation after the effects of an action have been projected. Here, I use the term *partial representation* because plan projection does not account for situational changes, which are exogenous to the agent's plan. For example, projecting situational changes associated with a plan for checking the mail would not account for changes that stem from the actions of other agents or those stemming from changes to the agent's environment.

During plan projection, the agent will project the situational changes associated with each action in the plan. A situational change occurs when an action introduces a new feature or removes or modifies an existing feature. For example, referring to Figure 7.4, exiting the house would result in a situational change because this action results in a change to the agent's location. Similarly, collecting the mail would result in a situational change because the action introduces a new feature (e.g. $Have(mail,agent,S_4)$) into the agent's situation.

Situational changes are predicted based on the agent's beliefs about a situation prior to projection and the postconditions it expects to be true after projection is complete. Therefore, the projection process assumes an action A_i is successfully executed in a situation S_i. The postconditions and non-effects associated with the action are then used to create a successor situation S_j, where $S_j = do(A_i, S_i)$. Here, *non-effects* mean that the frame problem is ignored and all features of a situation that were not modified by the agent's action are assumed to remain unchanged after the action is executed. For example, referring again to Figure 7.4, exiting the house has no effect on any situational features except for the agent's location. Therefore, it is assumed that agent's state (e.g. the clothes it is wearing) and the state of the stove will remain unchanged after the action exit house has been executed. Figure 7.4 shows the full situation history associated with the plan shown in Figure 7.3.

7.4.2.3. *Identifying future contexts*

The sequence of situation histories generated during plan projection provides an agent with a description of the situational changes its actions will induce during the problem-solving episode. This information can be used to identify any future contexts that will activate or alternatively deactivate as a result of those changes.

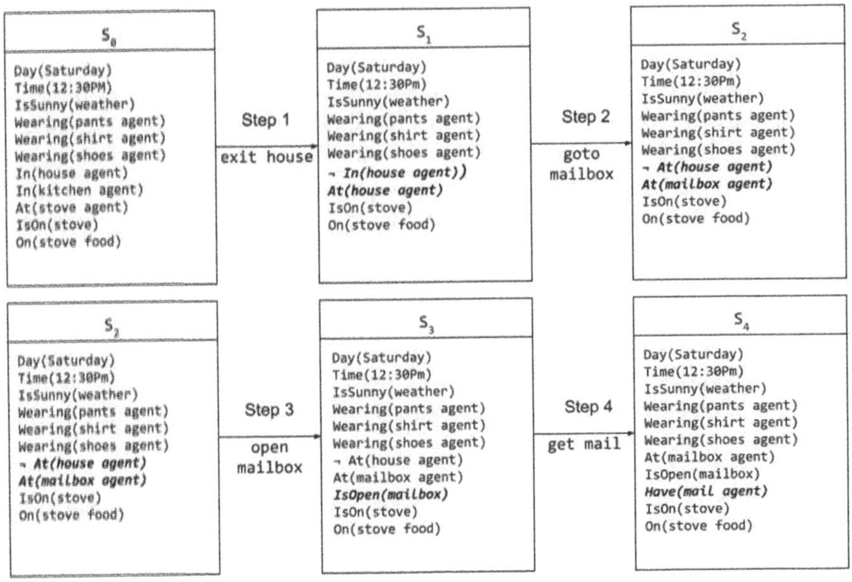

Figure 7.4. *Plan projection: check mail*

Similar to CMB, I treat context identification as a diagnostic process. After observing the features of a situation, an agent will compare them against the descriptive information in its c-schemas to determine if the situation contains an instance of a known context. This process is generalized to allow for future context identification across the sequence of situation histories.

To identify future contexts, an agent will apply the context identification process to each situation S in the sequence of situation histories. If a context is active in a situation S_j, where $S_i \sqsubset S_j$, and the context was not active in S_i, a reference to that situation is added to the problem-solving scenario. This reference, which I refer to as a context activation site (CAS) is a meta-descriptor that includes information about the context the agent believes will activate and, additionally, the action responsible for its activation. An example of a CAS is shown in Figure 7.5.

Sometimes, a context is defined in terms of multiple features and no single action is responsible for introducing all defining features into an agent's situation. In these cases, the features may be introduced across multiple actions. For these contexts, a CAS is declared when all defining features have been introduced into the agent's situation. Consequently, this means that a CAS will be declared for the successor situation of the action that introduces the last feature.

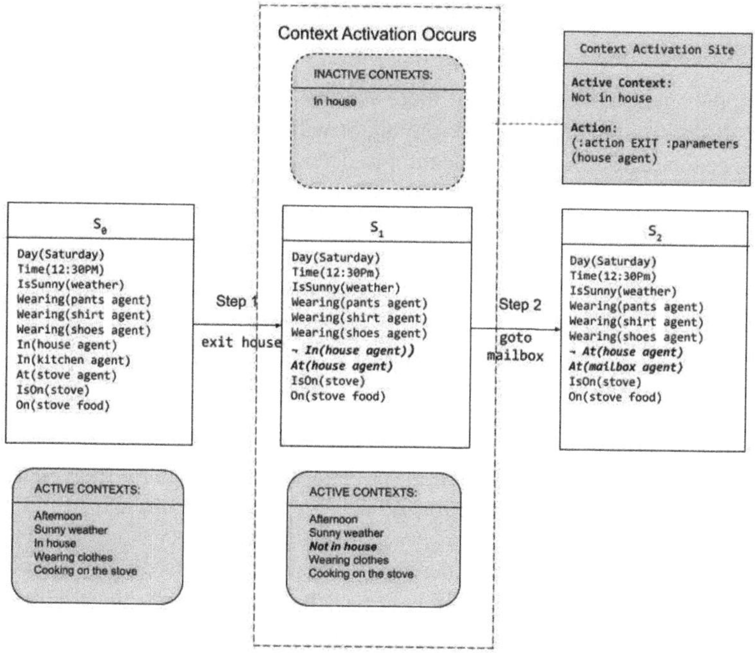

Figure 7.5. *Context activation site: check mail*

When taken together, context activation sites provide an agent with a temporally ordered description of the context transitions that will occur during the problem-solving episode.

7.4.3. *Identifying context-inappropriate behavior*

Once context prediction is complete, the agent will attempt to identify any context-inappropriate behavior related to the problem-solving scenario. This includes behaviors that are inappropriate for the agent's current context or any context it will encounter during the problem-solving episode.

In section 7.3.2.2, I introduced the concept of feature constraints, which are used to elicit context-appropriate behavior by placing implicit restrictions on the actions an agent can execute while it is in an instance of a particular context. Because feature constraints are used to elicit context-appropriate behavior, actions that cause the agent to violate those constraints are considered context-inappropriate. As such, the agent will use constraint violations as a primary indicator of context-inappropriate behavior.

First, the agent will determine if there are any constraints imposed on its current context. If constraints are imposed, the agent will search the sequence of situation histories generated during plan projection to determine if any of its projected actions will violate those constraints. For example, if the agent is currently in the context "cooking on the stove", feature constraints are imposed on the agent's location (see Figure 7.2). If a future situation reveals a change to the agent's location but the problematic feature(s) that define the constraints still exist, then a constraint violation is assumed to occur.

When a constraint violation is detected, a reference to the assumed violation is added to the problem-solving scenario. Similar to a context activation site, a constraint violation (CV) is a meta descriptor, which includes a reference to the constraint that is violated and the action (or actions) responsible for the violation. An example of a CSV is shown in Figure 7.6.

The agent will also attempt to identify constraint violations associated with each future context it expects to encounter. For each future context, the agent will use knowledge of that context to determine if the context has any associated constraints. If feature constraints are imposed, then the agent will search all successive situation histories for any constraint violations. If a violation is identified, a reference to the violation is added to the problem-solving scenario.

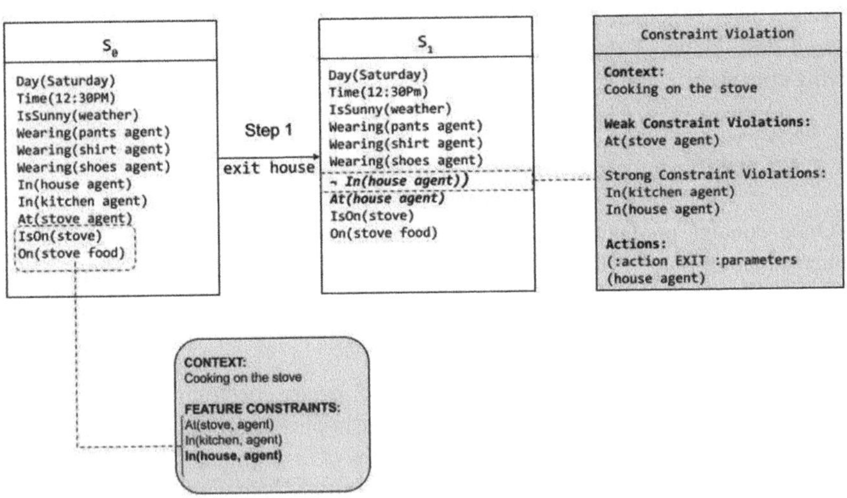

Figure 7.6. *Constraint violation: check mail*

7.4.4. *Strategy modification*

The final reasoning task in the CDBr framework allows the agent to effectively deal with any constraint violations (i.e. context-inappropriate behavior) it expects to occur during the problem-solving episode. Ideally, there are no constraints imposed on any context the agent will encounter or, alternatively, no imposed constraints are violated during the projected problem-solving episode. In this case, the agent can commit to the candidate goal using the chosen strategy.

However, if feature constraints are imposed and constraint violations are predicted, the agent should take steps to address any violations it predicted. As demonstrated in the previous section, constraint violations can occur when the agent's actions modify a constrained (i.e. non-problematic) feature but the constraining (i.e. problematic) feature still exists in the agent's situation. For example, the constraint violation shown in Figure 7.6 occurs because the agent's action modifies its location but the problematic features *IsOn*(*stove*) and *On*(*stovefood*) persist.

Certain types of agents such as automated planners might need to formulate a new plan in order to effectively deal with constraint violations. One advantage to the CDBr framework is that context prediction provides an agent with explicit information about constraint violations and the actions responsible for those violations. The agent can use this information to make targeted modifications to its plan, thus avoiding the need to completely replan. These modifications include augmenting a plan with:

1) specific actions to remove problematic features from the situation;

2) clean-up actions to remove situational features and deactivate incompatible contexts;

3) actions to modify contexts themselves.

An example of an augmented strategy for achieving the goal check mail is shown in Figure 7.7.

For some problem-solving scenarios, an agent may not be able to identify any actions or plan modifications that allow it to remedy any predicted constraint violations. When this happens, the agent can select or formulate an alternative plan for achieving the candidate goal. If the agent is unable to select or formulate an alternative plan, it might choose to defer the candidate goal until a later time.

```
Goal: Check mail

;; Agent is at the mailbox and possesses mail
Goal-state: (and (at (mailbox agent))
                 (have (mail agent))

;; Remove food from the hot burner
step:    (:action REMOVE :parameters (food burner)
          :preconditions (and
                   (at-stove agent)
                   (on-burner food)
          :postconditions (not (on-stove food)))
;; Turn of the burner
step:    (:action TURN-OFF :parameters (burner)
          :preconditions (and
                   (at-stove agent)
                   (is-on burner)
          :postconditions (not (on stove)))
;; Navigate to the front/main door
step-1:  (:action GOTO :parameters (door agent)
          :preconditions (in-house agent)
          :postconditions (at-door agent)))
;; Open the door
step-2:  (:action OPEN :parameters (front-door)
          :preconditions (at-door agent))
          :postconditions (is-open door))
;; Exit the house
step-3:  (:action EXIT :parameters (house agent)
          :preconditions (in-house agent)
          :postconditions (not (in-house agent))
;; Close the door
step-4:  (:action CLOSE :parameters (door agent)
          :preconditions (is-open door)
          :postconditions (not (is-open door))
;; Navigate to the mailbox
step-5:  (:action GOTO :parameters (mailbox agent)
          :preconditions (not (in-house agent))
          :postconditions (at (mailbox agent))
;; Open the mailbox
step-6:  (:action OPEN :parameters (mailbox agent)
          :preconditions (at (mailbox agent))
          :postconditions (is-open mailbox)
;; Get/obtain mail
step-7:  (:action GET :parameters (mail agent)
          :preconditions (and
                      (at (mailbox agent))
                      (is-open mailbox)
                      (contains (mail mailbox)))
          :postconditions have(mail agent)
;; Close the mailbox
step-8:  (:action CLOSE :parameters (mailbox agent)
          :preconditions (and (at (mailbox agent) (is-open mailbox))
          :postconditions (not (is-open mailbox))
```

Figure 7.7. *Augmented plan: check mail*

7.5. Conclusion

In this chapter, we have briefly discussed a framework for a proactive model of contextual reasoning. The framework allows for a novel form of contextual reasoning in which an agent predicts inappropriate behavior that can occur during plan execution and take steps to avoid such behaviors. One advantage of this approach is that it highlights specific steps that may result in inappropriate behaviors. This allows an agent to make context-specific modifications to a plan when it predicts that context-inappropriate behavior might occur, thus avoiding the need to completely replan. We intend to continue our work on the approach developed so far, using it to address limitations associated with other models of contextual reasoning. This includes using this approach both for virtual humans and for intelligent real-world agents.

7.6. References

Gonzalez, A.J. and Ahlers, R. (1998). Context-based representation of intelligent behavior in training simulations. *Transactions of the Society for Computer Simulation*, 15(4), 153–166.

Hanks, S.J. (1991). Projecting plans for uncertain worlds. Doctoral Dissertation, Yale University, New Haven, CT.

Levesque, H., Pirri, F., Reiter, R. (1998). Foundations for the situation calculus. *Electronic Transactions on Artificial Intelligence*, 2(3–4), 159–178 [Online]. Available at: http://www.ep.liu.se/ej/etai/1998/005/.

McCarthy, J. (1963). Situations, actions, and causal laws. Technical Report, Stanford University, Department of Computer Science, CA.

McDermott, D. (1997). Probabilistic projection in planning. In *Spatial and Temporal Reasoning*, Stock, O. (ed.). Springer, Dordrecht.

Turner, R.M. (1994). *Adaptive Reasoning For Real-World Problems: A Schema-Based Approach*. Psychology Press, Hove.

Turner, R.M. (1998). Context-mediated behavior for intelligent agents. *International Journal of Human–Computer Studies*, 48(3), 307–330.

Turner, R.M. (1999). Context-mediated behavior: An approach to explicitly representing contexts and contextual knowledge for AI applications. AAAI Technical Report [Online]. Available at: https://www.aaai.org/Papers/Workshops/1999/WS-99-14/WS99-14-017.pdf.

Wilson, C. and Turner, R.M. (2017). Context and the virtual human. *Proceedings of the 10th International and Interdisciplinary Conference on Modeling and Using Context*, 227–239. Roskilde.

8

Context-Based Personal Data Discovery for Anonymization

The rapid technological change and globalization have created new challenges for the protection of personal data, with the scale of their collection and sharing increasing exponentially. The anonymization of data is one of the methods of complying with the General Data Protection Regulation (GDPR). Protecting sensitive business data, and in particular customer personal data, starts with knowing what sensitive data is and where it is located. Data discovery inspects metadata and real data in different databases to discover personal data, and provides comprehensive results listing the sensitive fields storing that data. The specification of personal data is a very time-consuming phase for business experts in the analysis, especially in the case of databases with a large amount of data stored in thousands of fields in the tables. Automatic data discovery tools cannot always extract the exhaustive list of personal data and may incorrectly specify certain data as personal. The error deviation in automatic detection is due to the context in which the data is used. This chapter presents how to consider context in personal data discovery to improve the anonymization process. We use the contextual graphs formalism to improve procedures of sensitive personal data discovery used in the anonymization process.

8.1. Introduction

On a daily basis, billions of pieces of data leave production environments without being protected: copies of production for developers and administrators, new features and analytical models developed on an environment close to production. Personal data is a valuable resource for many organizations, e-commerce companies and social media networks. Data sources are highly coveted, and can be consulted and used by

---

Chapter written by Hassane TAHIR and Patrick BRÉZILLON.

various actors in economic life. However, in the digital age, they are unfortunately also subject to illegal treatment and numerous misappropriations without the knowledge of the people from whom they emanate. The GDPR (General Data Protection Regulation) has been established as a legal framework so that each piece of data is used wisely and protected (Intersoft 2021). The main objective is to reassure and protect Internet users, particularly when connecting to social networks and e-commerce applications.

The exposure of sensitive data to non-production environments (i.e. development and test) can create serious problems. Some types of information, such as customers' names, addresses, emails and credit card numbers, are inherently sensitive. They always need to be kept out of production environments, and any database that holds them needs protection. Protecting sensitive business data, and in particular customer personal data, starts with knowing what sensitive data is and where it is located. Data discovery inspects metadata and real data in different databases to discover personal data, and provides comprehensive results listing sensitive fields storing such data (Oracle 2021). Humans can distinguish different contexts in the classification of data. The name of a person by itself is not sensitive data. Nonetheless, a data analyst can distinguish between different contexts where an individual's name may be moderately sensitive or highly confidential. For example, in the case of personal information in medical forms and records, merely disclosing just the patient's name may constitute a significant violation. The process of data discovery can be improved by different contextualizations of connected data groups through user practices added to well-established procedures. For instance, a field labeled "first name" will not be considered as a field containing personal data if all the records in the database for this field always return the values "Zoe" and "Megane", which are certainly the first names of people, but in the application context, correspond to names given to categories of car models.

This chapter presents how to consider context in the procedures of personal data discovery to improve the anonymization process adopted by a company to protect its corporate applications and those of its international subsidiaries. We use the contextual graphs formalism (Brézillon 2018) as a support tool to manage all the contexts involved in the process of personal data discovery. In this way, practices replace procedures by contextualized procedures. Moreover, the addition of practices, which are performed by business analysts, to the existing set of procedures in an incremental way leads us to build an experience discovery database useful for the panel of involved actors (business analysts, product owners, security managers and anonymization developers) who share their knowledge on the classification of sensitive data. This chapter is organized as follows. Section 7.2 introduces personal and sensitive data. Section 7.3 presents the procedure of personal data discovery (objective of personal data discovery procedures, role of a DPO (data protection officer) in personal data discovery, description of procedure of data discovery).

Section 7.4 shows how to specify personal data in the context of an anonymization process (definition of anonymization, motivation for data anonymization, examples of techniques of anonymization, anonymization process, contextual elements in personal data discovery). Section 7.5 summarizes some related work. Section 7.6 presents how to contextualize the procedure for data discovery (the concept of context, Conceptual Graph Approach, a case study). A conclusion of the present work ends this chapter.

8.2. Personal and sensitive data

Personal data is defined by the CNIL (French regulatory institution for personal data) as: "Any information relating to an identified or identifiable individual; an identifiable person is one who can be identified, directly (i.e. last name, first name) or indirectly (i.e. phone number, license plate of a vehicle, an identifier such as social security number, postal code or email address, but also voice or image)" (CNIL 2021). The identification of a person can be carried out using a single piece of data (i.e. name) or a data correlation (i.e. a woman living at a given address, born on a given day and being a member of a given association). On the contrary, company contact details (i.e. the company "Company ZYX" with its postal address, the telephone number of its switchboard and a generic contact email "company_xyx@email.fr") are not, in principle, personal data.

Sensitive data is a special category of personal data that is related to racial or ethnic origin, political opinions, religious or philosophical beliefs or trade union membership, as well as genetic data, biometric data for the purpose of uniquely identifying a person or an individual, data regarding health, sexual life or orientation of a person. The European regulation prohibits the collection or use of this data, except in the following cases, particularly:

– if an individual has given their express consent (active, explicit and preferably in writing, which must be free, specific and informed);

– if the information is made public by an individual;

– if the data is required for the safety of human life;

– if the data is authorized by the CNIL and the reason for using this data is of public interest;

– if the data is about members of a political, religious, philosophical, political association or organization, or trade union.

Table 8.1 presents a non-exhaustive list of personal data, where we use only the term of personal data.

Personal Data Categories	List of Personal Data
Identifying Data	Last Name and Surname
	Email address
	Telephone number
	Mailing address
	Photo
	Birth date / Age
	Nationality
	Family status (marital status, children)
	ID data (social security number, ID number, etc.)
Location data	GPS coordinates, City, postal code, Street name
Professional life / Education	Company position, Company name, Company ID number
	Professional coordinates (email address, phone, company mailing address)
Banking data	Banking ID (account number, RIB, IBAN, SWIFT)
Internet (Connexion) data	IP address, Host names
Medical data	Physical characteristics
	Health status (sickness, allergies, etc.)
	Biometric data
Beliefs	Political beliefs
	Philosophical beliefs
	Religious beliefs
Recordings	Video recordings
	Sound recordings
Car Identification	VIN number
	Vehicle registration plate (Car number plate, license plate)

Table 8.1. *A non-exhaustive list of personal data*

8.3. Procedure of personal data discovery

8.3.1. *Objective of personal data discovery procedures*

Personal data discovery is the most important activity in the data analysis stage of the anonymization process. From a technical standpoint, the easiest way is to have someone examine all database sources for personal/sensitive data discovery and classification. This can generate accurate results, with few false positives, but getting it right is problematic. However, someone assigned to review many database tables can become careless and miss some data that should be flagged. People are often bad at performing repetitive and mundane tasks.

Data discovery procedures have been established to automate repetitive actions to detect personal data based on different criteria and techniques, such as using regular expressions and rules based on dictionaries of cities, first names, etc. Once the discovery procedure is set up, it has many other uses besides data security and privacy compliance. By applying a different set of rules, it can help us to locate relevant information in the archives.

8.3.2. *Role of a DPO in personal data discovery*

Many actors can be involved in setting up personal data discovery procedures, such as data protection officers (DPO), project managers, business data analysts, developers and database administrators. The main actor is the DPO because they can be easily accessible to anyone at an organization to oversee data protection. A DPO must have skills in data protection law and keep up with evolving legislation. They should have a complete understanding of the company's IT infrastructure, technology and a techno-organizational structure. The DPO expertise should align with the organization's data processing operations and the level of data protection required for what is processed by data controllers and data processors.

8.3.3. *Description of procedure of data discovery*

Data discovery procedures use sensitive types that define the types of data to look for. Sensitive types cover personal data relating to sensitive personal data, including information related to racial or ethnic origin, political opinions, religious or philosophical beliefs, financial data, information concerning the health or sex life of an individual. The predefined sensitive types can be created and organized into categories, making it easier to find and use the relevant sensitive types. Data discovery can be used to detect and find sensitive columns in database tables that meet specified criteria. The discovery procedure is composed of three steps. The first step specifies the target database and schemas where personal data is located. The next step identifies sensitive columns by examining column names, comments, data samples and object relations. Finally, a report of sensitive data is generated to be reviewed by data analysts. Figure 8.1 shows the type of fields detected by the discovery tool that store personal data.

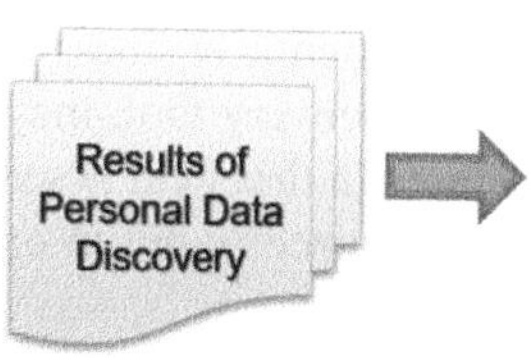

List of Personal Data	Fields storing Personal Data
Last Name	LAST_NAME_1
	LAST_NAME_2
	COMPANYNAME
First Name	FIRST_NAME_1
	FIRST_NAME_2
	TEXT_2
City	CITY_NAME_1
	TEXT_3

Figure 8.1. *Example of fields with personal data detected by a discovery tool*

Data discovery procedures are widely used by tools for the automatic detection of personal data. In our research, we have used a tool developed in Python, which can be connected to database sources such as Oracle, IBM DB2 and SQL Server. The objective is to analyze data stored in the different tables of a database, in order to determine the fields containing personal data using regular expressions and dictionaries of last and first names of people, city names, etc. In the platform of a data discovery tool, the anonymization developer requests that the personal data is read from the production source database and displayed in a suitable form for users. Table 8.2 shows an example of a result sheet with a classification of personal data and how many times each piece of data is found in the database.

Column_name	Location	Numbers_of_rows_in_table	Numbers_of_rows_analyzed	Percentage_of_personnal_data	Data_found_count
birth_date	BIRTHDATE	701389	1000	56.3	563
lastname	COMPANYNAME	701389	1000	2.4	24
e_mail	EMAIL	701389	1000	59.8	598
firstname	FIRSTNAME	701389	1000	35.2	352
lastname	FIRSTNAME	701389	1000	6.2	62
town	FIRSTNAME	701389	1000	2.2	22
town	HOMECITY	701389	1000	2.4	24
firstname	LASTNAME	701389	1000	1.0	10
lastname	LASTNAME	701389	1000	8.6	86
town	LASTNAME	701389	1000	1.3	13
firstname	TEXT	610	610	5.15	14

Table 8.2. *An example of the results of personal data detected by a discovery tool*

The discovery tool cannot always perform an accurate classification of personal data. For example, a field named "TEXT_2" may store names classified by a tool as first names (i.e. personal data), but may also store names of cars (i.e. Zoe, Mégane) which should not be considered as personal data. Therefore, the context regarding the data (i.e. contents) should be considered in the procedure of data discovery.

An automatic data discovery tool classifies fields in a database as personal data such as last name, first name, email, town (or city). But when a business analyst examines the result of personal data classification by the discovery tool, they must check each field or column to determine whether data stored in that field is sensitive personal data or not, and therefore whether it will be anonymized or not in order to be GDPR compliant. Figure 8.2 shows areas with problems.

List of Personal Data	Fields storing personal data
Last Name	LAST_NAME_1
	LAST_NAME_2
	COMPANYNAME→ not a Last Name ('HAMILTON', 'JOHNSON')
First Name	FIRST_NAME_1
	FIRST_NAME_2
	TEXT_1→ add the field TEXT_1 to the list of personal data
	TEXT_2 → not a First Name ('ZOE','MEGANE')
City	CITY_NAME_1
	TEXT_3
	CITY_NAME_2 → This field is empty but can be filled with city names.

Figure 8.2. *Example of problem detection by the discovery tool*

More formally, Table 8.3 explains how the business analyst contextualizes the results of personal data discovery.

List of Personal Data	List of fields storing Personal data	Discovery Tool Result: Fields storing Personal data?	Data Analyst Investigation: Fields storing Personal data?	Comment
Last Name	LAST_NAME_1	Yes	Yes	
	LAST_NAME_2	Yes	Yes	
	COMPANYNAME	Yes	No	The field COMPANYNAME is not a Last Name even if it stores the values such as 'HAMILTON', 'JOHNSON'.
First Name	FIRST_NAME_1	Yes	Yes	
	FIRST_NAME_2	Yes	Yes	
	TEXT_2	Yes	No	The field TEXT_2 is not a First Name even if it stores the values such as 'ZOE','MEGANE'
	TEXT_1	No	Yes	The Analyst confirms that the field TEXT_1 is a First Name. It is not detected by the discovery tool as a First Name because it is empty.
City	CITY_NAME_1	Yes	Yes	
	TEXT_3	Yes	Yes	
	CITY_NAME_2	No	Yes	The Analyst confirms that the field CITY_NAME_2 is a city. It is not detected by the discovery tool as a city because it is empty.

Table 8.3. *Contextualization of the results of data discovery*

8.4. Specifying personal data in the context of an anonymization process

8.4.1. *Definition of anonymization*

Anonymization is a technique that irreversibly prevents the identification of an individual's personal data. European countries and the USA have strengthened personal data protection by developing comprehensive procedures that give people more control over their own data. It is also considered as an operation that consists of hiding or replacing information (personal or other data) of different types and origins before sending it to third parties (as with the open data movement, providers, etc.). Clearly, information can be replaced with values that no longer carry any information, as these values do not make it possible to find the original information. Therefore, anonymization can be applied to all information that makes it possible to identify a natural person directly or indirectly, including by cross-checking or cross-referencing with other information from the same source, or from other sources (even external). Typical examples are first name, last name, street, cities, age, date of birth, telephone numbers, email address, etc. Other terms used to refer to anonymization include data masking, data cloaking and data masquerading.

On a specific IT project, it is therefore necessary to analyze and identify all the information that will have to be anonymized in a precise and unambiguous way. This task, which falls to the project data manager (usually the business analyst), is crucial because the slightest oversight can have unfortunate consequences: once the data is outsourced, it will be too late. But why do we need to anonymize?

8.4.2. *Motivation for data anonymization*

The most common motivation of data anonymization is the protection of personal data as adopted by the law of Informatics and Freedom ("Informatique et Libertés") of January 6, 1978, which is at the origin of the creation of the French CNIL Organization. The second motivation is the protection of professional and ethical secrecy. The third reason is the protection of financial and economic information. Other motivations include the protection of commercial and/or contractual information, as well as the protection of strategic information (with a highly competitive value).

In general, anonymization is useful in all areas where confidential information about individuals must be shared with third parties. For example, in the medical field, it is not uncommon for patient data to be transmitted to research institutes. Another common case of the use of anonymization techniques is the sharing of data regarding a government's citizens. The latter uses private companies for the development of its software, and in this case, it must provide its partner with test data. Databases containing highly sensitive data should ideally be transmitted anonymously.

8.4.3. *Examples of techniques of anonymization*

Many techniques are presented in the literature. Randomization and generalization are two anonymization approaches. The first technique of the randomization procedure is "noise addition", such as changing a height of a person from 178 cm to 165 cm. The second technique is substitution/permutation, such as replacing a zip code 92600 by 32000. In a generalization procedure, personal identifiers are generalized into a range or group, for instance by replacing an age of 45 years by an age of 40–50 years, using techniques such as l-diversity and aggregation/k-anonymity.

There are many anonymization techniques, and each of them applies to particular uses.

1) Replace the original real source data with unusable data

This technique consists of adding noise to the data, in order to drown the information in the mass. For example, if we want to anonymize the age of a patient from 32 years, we could replace this information with a range, such as 28–35 years. It then becomes impossible to find the exact age of the person. It is also possible to play on numerical values and dates by changing them by a certain percentage.

2) Only keep useful data

Another simple but powerful technique is simply to delete the data. Obviously, this does not apply to all situations. This technique can be used in cases where certain information is not useful to a third-party partner. Deleting data prevents any attack that could lead to a data leak.

3) Anonymize data by substitution

It is possible to anonymize by substitution. The idea of this technique is to replace the existing data with other data that is similar to the original. If we take the example where a government administration sends a database of individuals to a third-party company for software development, people's names should never be revealed. To enable developers to work and rely on consistent data, the personal data is then substituted by random values. For instance, BERGER Jacques is replaced by THOMAS Arthur.

Figure 8.3 shows an example of the above-mentioned three anonymization techniques (1, 2 and 3) applied to an SQL database.

Last Name	First Name	Age	Social Security Number
BERGER	Jacques	63	157042531111426
SIMON	Nathalie	26	294037512000522
DUBOIS	Philippe	51	169054958815780
PETIT	Alain	35	185029510501260
MULLER	Lara	27	293119231111426

Last Name	First Name	Age	Social Security Number
THOMAS	Arthur	60+	*
MOREAU	Brigitte	25-30	*
BOUCHER	Paul	48-55	*
RICHARD	Pierre	30-40	*
CARPENTIER	Elisabeth	25-35	*

Figure 8.3. *An example of the anonymization technique*

In the first table, we have many users, where their last name, first name, age and social security number are present in the plain text. In the second table, we find this anonymized personal data as follows:

– last name: this is replaced by a random last name;

– first name: this is replaced by a random first name;

– age: this is substituted by an age interval in order to hide the actual age;

– social security number: this has simply been deleted (i.e. removed).

All of the techniques presented above are examples of randomization. We try to reduce the relation and links between a person and their data in order to make them unidentifiable.

4) Drowning data through generalization

There are also methods using generalization, that is, drowning information in the mass in order to reduce the possibility of identifying a person by crossing several sources of information. There are several methods, including k-anonymity and l-diversity.

8.4.4. *Anonymization process*

A large amount of data is produced in production environments every day without being protected. According to Oracle, the data to be anonymized represents 75% of the non-production environment (i.e. the test environment). The GDPR has led companies to strengthen their security in order to protect their data. It clarifies that anonymized data is no longer subject to the regulation. Therefore, each data manager must anonymize certain data so that it is used for justified and limited purposes. The simplest approach to anonymizing data may be to identify data fields that store personal information (i.e. social security numbers, email and IP addresses). Privacy needs can be general and above all specific to the system under consideration.

Implementation is perhaps the greatest difficulty that characterizes the data anonymization process. In fact, it can raise several questions about the context of anonymization. Should I anonymize by application? What management should be adopted for applications sharing personal data? Which mode of organization will better meet the business requirements? These are some of the important questions that testify to the difficulty of setting up this technique. This means that organizing the anonymization process is the essential strategic axis that determines its success. For this, it would be useful to set up an industrialized service intended to anonymize and meet the needs of the IT teams most concerned. This involves an ability to address all technologies and a powerful and intelligent sampling proposal, based on a large dataset in data sources and repositories and setting up a library of formats to anonymize completely (random substitution, deletion, rewriting) that may guarantee efficient service levels. This service will then improve the working methods of IT teams, without negatively impacting their daily lives.

Under the GDPR, anonymizing data has the same effectiveness as deleting it. Data deletion may concern a single attribute such as the right of an individual to be forgotten with respect to their e-mail address. In contrast to data deletion, anonymization relates to all data regarding a person. Being irreversible, the anonymization technique cannot be applied to data in the production environment that is used by business processes, as these would be unusable. What would suggest anonymization rather than data deletion is the use of that data in the context of statistical analysis (e.g. for marketing purposes) and testing in the software development cycle (unit tests, integration, qualification). There are many data anonymization techniques proposed to comply with the GDPR, which depend on the type of data to be processed, as well as on how the data will be used.

Anonymization is a technique applied to personal data to irreversibly prevent their identification. The starting point is therefore that the personal data must have been collected and processed in compliance with the applicable legislation on data retention in an identifiable form. The anonymization process can be divided into four main phases, which can be summarized as follows. The first phase is the data discovery. Sensitive data must be discovered, and inter-table relationships must be specified. The second phase is to define masking rules and/or the subsets (or the sampling criteria). The third phase corresponds to testing and validation of the anonymization. The last phase is the industrialization and execution of the anonymization to publish masked data. The essential steps in an anonymization project are as follows:

1) start by identifying which data sources need to be anonymized (database(s), incoming and/or outgoing data flows, document template(s), document(s), etc.);

2) choose whether anonymization should be performed on the fly or in bulk for each data source. The first (on the fly) is complex, the second (in bulk) is the most common case at the database level;

3) identify and precisely locate (for each source) all the data to be anonymized;

4) define, according to each type (business) of data (date, last name/first name, address, telephone number, city, etc.), the anonymization technique to be used (see above), and the business and technical constraints that should be respected;

5) choose a technical solution that makes it possible to implement all of the defined anonymization rules;

6) implement the anonymization process (development of a specific tool, or configuration and development using a software solution from a provider);

7) perform an in-depth procedure of testing and validation of the anonymization with data close to that of production. This is not a usual recipe of a classic IT project: the slightest omission of a data (not anonymized), or a poor-quality anonymization (i.e. reversible process), will be irreversible once in production.

This anonymization development process can obviously be carried out in an iterative and agile way.

8.4.5. *Contextual elements in personal data discovery*

Generally, all companies process personal data, particularly that of their employees. To know the location of the personal data that the company holds, we need to have a clear overview of the data and where it is located. It is also necessary to identify these data and record them in order to link them to a treatment. Two approaches can be combined for greater efficiency: "bottom-up" and/or "top-down" approach. In the top-down approach, we must look at the structure of the company, its departments, its sub-departments and what personal data are processed by these departments. It is also necessary to meet key people to better understand personal data and their management. The company's organizational chart makes it possible to identify key people and discern the contours of the main categories of data processing. Management support is essential. The stakes must be understood by the business leader and managers so that the necessary resources for the proper execution of the process are made available. In the bottom-up approach, the structure of the computer network, the database and its tree structure are generally good indicators. It is important to ask IT providers for support in establishing the tree structure of the information processing carried out within the company required for data discovery to constitute an accurate map.

In the process of data discovery, many data, especially personal data, are not classified or linked to a specific process. It is therefore appropriate to classify them, or link them to the corresponding data process. It is also necessary to distinguish between external personal data (i.e. customers, suppliers, prospects) and internal personal data (i.e. employees of the company). This will make it possible to determine who is responsible (data controller, supplier). Therefore, the identification of data and their assignment to any processing makes it possible to delete all personal data that have become obsolete or useless and only keep what is useful and necessary. The accumulation of personal data is prohibited by the regulation. For this reason, it is important to manage them adequately and erase them when they are obsolete or when the processing for which they have been collected is completed.

Identifying personal data is an important phase of a data anonymization process. This is not always the easiest activity, particularly when data sources are not clearly defined or well documented. Data sources may also belong to an old technology, which may complicate the analysis by business managers who may not have the required knowledge about the databases used. In the context of data anonymization, the first task is to gather the required information about the contextual elements regarding data in the different applications of an organization or corporate company. For these reasons, data analysts use automatic data discovery tools which are very useful in supporting them to specify sensitive and personal data. Nevertheless, such tools do not always retrieve the required personal data that should be anonymized to comply with the GDPR. Many fields may be personal data to store names or first names, but if they are empty, the discovery procedure cannot detect them. Another example concerns fields that are code or a functional key that may contain business information. In a code JA870612341, JA are the first two characters of the lastname JAMES, followed by the year and month of birth. The last character stands for the gender. Such a code cannot be detected by discovery tools.

Type of context	List of contextual elements	Values of contextual elements
Organizational context	Business area	(HR, Finance, Purchasing, etc.)
	Application managers	Technical Manager, Business Manageer, Data Analysts, DBA
	Level of business or functional knowledge of the application	(1 = none, 2 = medium or 3 = good)
	Level of knowledge of the data model	(1 = none, 2 = average or 3 = good)
	Perenniality	Ongoing, 6 months, 1 year, 2 years, 3 years, unknown
Functional/business context	Quantity of personal data (approximately)	0–50 fields, 50–100 fields, more than 100 fields, unknown
	Does application contain HR data?	Yes, no, unknown
	Does application contain customer data?	Yes, no, unknown
	Does application contain supplier data?	Yes, no, unknown
	Presence of sensitive data	Yes, no, unknown
	Type of sensitive data present in the application	Racial or ethnic origin, political, philosophical or religious opinions, trade union membership, health or sexual life
	Is it a referential application that centralizes all customer data?	Yes, no, unknown
Technical context	Number of environments out of production (test, dev, recept, etc.) loaded from production environment	0,1,2,3,...,unknown
	Number of environments out of production (test, dev, recept, etc.) loaded from non-production environment	0,1,2,3,...,unknown
	DBMS and versions	DB2, Oracle 11, Oracle 12, SQL Server 2012, MySQL 8, Sybase, unknown
	Platform, OS and version	Unix, Linux, Windows 10, MainFrame, unknown
	Volume of the database in GB	<1Go, <10Go, <100Go, <200Go, <300Go,...,unknown
	Volume of the database in number of tables	<10 tables, <50 tables, <100 tables, <200 tables,...,unknown
	Cloud application?	Not a cloud, Amazon, Microsoft, unknown
	Host (customer or other?)	Unknown host, host name

Table 8.4. *Contexts of data discovery*

Three types of contexts can be distinguished when identifying data in the different applications or information systems, which are organizational context, functional or business context, and technical context. Each of these three contexts includes a list of related contextual elements. Examples of the values of each contextual element are presented in Table 8.4. Our study also considers other types of contexts about the data included in the different data sources.

8.5. Related work

Capturing sensitive personal data requires specific software tools to avoid costly disclosures and wasted resources. Approaches can range from purely manual to highly automated, although some human intervention is always required. The objective is to comply with common industry regulatory standards for data protection and privacy including the GDPR and other important regulations. The following are some related research works. Securiti (2021) is an artificial intelligence-based tool that helps organizations to discover sensitive data stored in different platforms (cloud, SaaS and on-premises environments), protecting that data and automating many of the privacy functions. Nightfall (2021) is a machine learning-based tool that supports a business company to manage and protect critical information such as personal data about customers, through data infrastructure, SaaS and APIs. IBM Security Guardium (IBM 2021) is a tool used to analyze, discover, classify and control access to sensitive data. The discussion on other related research work can be found in Abdalla (2013), Vianna et al. (2014), Kudim and Yu (2019), Siteimprove (2020) and Microsoft (2021). Finally, it is important to mention some anonymization tools and approaches to anonymize data such as the SMD Anonymizer (Santos et al. 2018), developed to generate anonymized files using the k-anonymity algorithm to improve data privacy. Prasser et al. (2014) propose an anonymization tool named ARX that can efficiently implement various privacy techniques and thus create an intuitive multi-platform graphical interface. The discussion on other related research work can be found in Bezzi et al. (2014), Nassar et al. (2015), Bawar et al. (2018), Vardalachakis et al. (2019), Elliot et al. (2020) and SoftwareTestingHelp (2021).

The above-mentioned solutions cannot always accurately specify all of the sensitive personal data that need to be anonymized. Up until now, the improvement of the existing data discovery procedures has been achieved through business analysts' practices that adapt these procedures to the context in which the data should be considered as personal data. Discovery tools cannot always extract the exhaustive list of personal data or may incorrectly specify certain data as sensitive. The error deviation in automatic detection is due to the context in which the data is used. For example, strings such as "username" and "password" can indicate sensitive personal data, but often this is not the case. The automated software tools (Bitterman 2019) may exclude uses whose context makes them probably safe. They can exclude some patterns (i.e. dummy strings such as all zeros) from being flagged as sensitive, thereby reducing the number of false positives. Better results may come from software tools designed to categorize sensitive data in a contextual way (Pretty 2020; Yayik et al. 2021). This covers a range of approaches.

8.6. Procedure contextualization for data discovery

This section presents features of the proposed approach for representing business data analysts' practices as contextualization of procedures.

8.6.1. *The concept of context*

The concept of context can help to address dynamic change and requirements in data discovery procedures. Brézillon and Pomerol (1999) defined context as "what constrains something without intervening in it explicitly". Three categories of context can be mentioned, namely external knowledge, contextual knowledge and proceduralized context. External knowledge is knowledge that has no relation with the current focus. Contextual knowledge is knowledge that is more or less related to the current focus. Proceduralized context involves the actor, who proceduralizes a part of the contextual knowledge to address the current focus. Our study focuses both on technical and user (or human) contexts. The technical context is related to the knowledge about the structure of data, the environment of the database containing data (production, test, etc.), data flow, security tools used to protect data, etc. The user context concerns human (business data analyst) knowledge and specific conditions to consider when classifying personal data that should be anonymized in the information systems of an organization.

8.6.2. *Conceptual graph approach*

The basic components of a contextual graph (CxG) are actions and contextual elements. Activities and Executive Structures of Independent Activities are two other elements of a CxG. An action can be represented on one or more paths, but it will be in different contexts. A contextual graph has the following properties: 1) it is a directed acyclic graph, 2) it has one input, 3) it has one output and 4) it has a spindle structure. It can model the different ways to solve a problem. A user practice can be represented in each path of a CxG, in order to show how an action is performed in each context. Brézillon (2005) and Tahir and Brézillon (2011, 2013) present more details about the CxG formalism, as well as the way it is implemented.

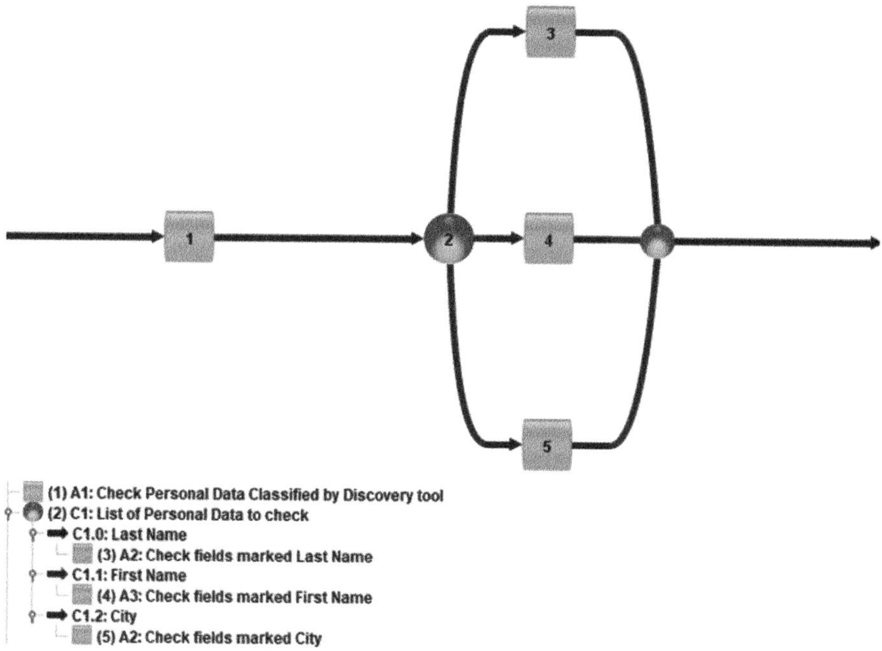

Figure 8.4. *Contextual graphs representing a simple procedure of data discovery. For a color version of this figure, see www.iste.co.uk/brezillon/context.zip*

Figure 8.4 shows a contextual graph representing a procedure of data discovery, where a business analyst checks personal data detected and classified by the discovery tool in order to adjust and contextualize the automatic detection of the data. The activity is also a CxG and can appear on different paths (i.e. different contexts) of contextual graphs (as in the case of actions).

A contextual graph is a knowledge acquisition tool (KAT) that has been improved to generate the user's practices data, which will be useful in creating indicators that can be used to assist the user in reading practices included in a given contextual graph. The main components of the KAT contextual graph (see Figure 8.5) are CxG Editor, CxG Reader and an experience database. The CxG Editor enables authorized users to manage their corresponding contextual graphs representing the main procedures and the significant changes added by them (i.e. practices). The CxG Reader makes it possible to read only a desired contextual graph to execute one or more practices already created by different data analyst experts to perform a given task or activity. The reading process is enabled when the user activates the corresponding CxG Reader module included in the KAT contextual graph menu. An experience database is used to record and store users' procedures and practices. This database stores all information about the contextual graphs objects created by data analysts in the form of an XML Schema. A path (Pi) corresponds to the user's practice, which is defined using two main sets: ordered actions (Ai) and contextual elements values (Cij). All of the relevant user's practices are saved in files that will be used as sources to load useful and relevant data into an operational experience database.

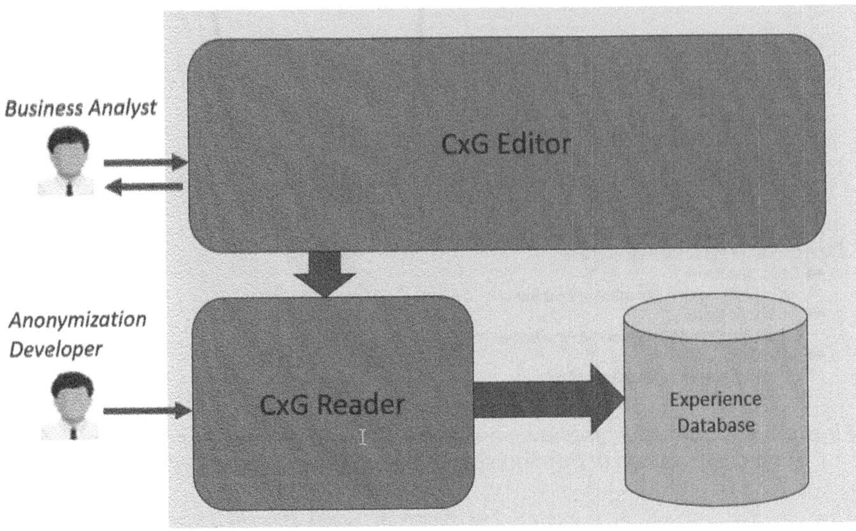

Figure 8.5. *CxG KAT (Knowledge Acquisition Tool) platform*

8.6.3. *A case study*

Contextual graphs software can be used as a decision-making tool to assist business analysts in personal data classification when working on databases with too many tables. Figure 8.6 presents a contextual graph used by a business analyst to adapt the result classification of personal data (by the discovery tool) to the context about the meaning of the data stored in each field of the database table. For instance, a contextual element (node 4) is used to determine whether the field named "COMPANYNAME" is really personal data storing "Last Names", as listed by the personal data discovery tool. Another example is the contextual element (node 9) used to verify whether the field named "TEXT_2" is really personal data storing

"First Names". The analyst removes the fields "COMPANYNAME", "TEXT_2" from the list of personal data and sets them to N/A (Not Applicable). Note that as indicated in the branch C9.1, the analyst adds a new field "TEXT_1" that is not detected by the automatic personal data discovery tool.

Sometimes the analyst may cooperate with other experts to confirm the status of other sensitive personal fields. The analyst interacts with the data project manager to share the same context about the personal data to check for some fields, as is the case of the field "TEXT_3" illustrated in branch C9.1 of the contextual graph. If this field often contains First Names (70% in scanned rows), the project manager may confirm a "TEXT_3" field as a First Name, which is personal data. But if this field contains names of cars (e.g. ZOE, which is a name of a Renault car), this field is not considered as personal data by the analyst. Contextual elements must be distinguished from their instantiations. It would not be enough to give the instantiations for a person/company. Two contextual elements are considered relevant together based on their instantiations.

A contextual graph is composed of many paths. A path (Pi) corresponds to the user's practice, which is defined using two main sets: ordered actions (Ai) and contextual elements values (Cij). All of the relevant user's practices are saved in the files that will be used as sources to load useful and relevant data into an operational experience database. This CxG is composed of the following:

– Four contextual elements C2, C4, C9 and C17 representing, respectively, nodes numbered 2, 4, 9 and 17 with the set of values:

Val (C2) = {C2.1, C2.2, C2.3},

Val (C4) = {C4.1, C4.2},

Val (C9) = {C9.1, C9.2, C9.3},

Val (C17) = {C17.1, C17.2}.

– Set of actions A = {A1, A2, A4, A5, A6, A7, A8, A9, A10, A12, A13, A14, A15, A16, A17, A18, A19}.

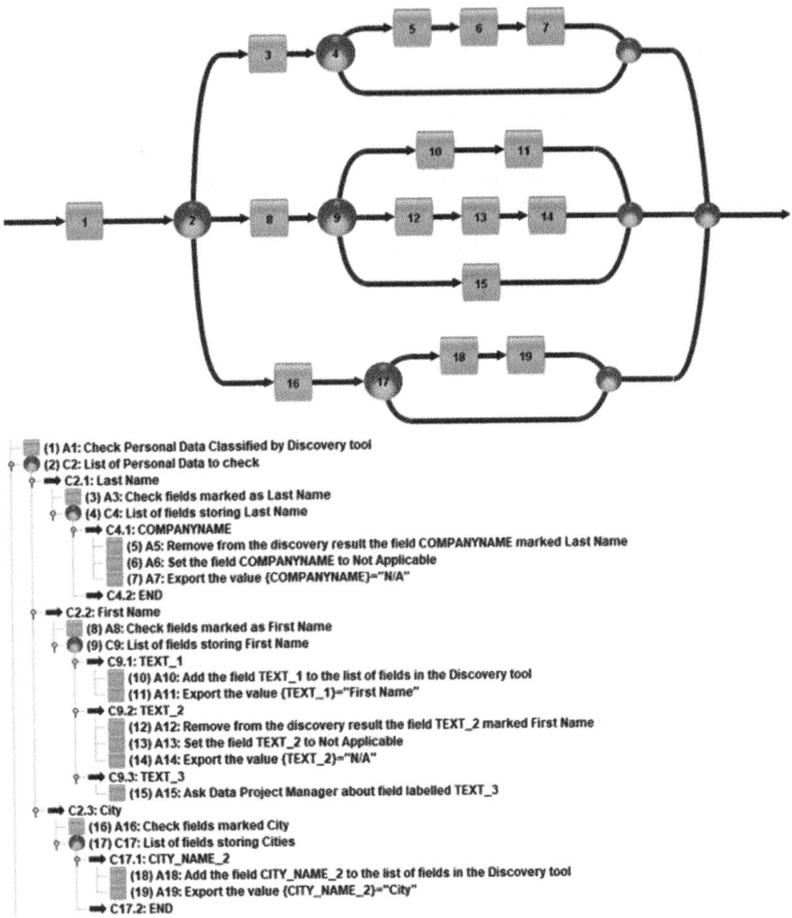

Figure 8.6. *Contextual graph assisting users in specifying personal data. For a color version of this figure, see www.iste.co.uk/brezillon/context.zip*

The labels of the analyst actions, contextual elements and their values are indicated in Figure 8.6. When reading a contextual graph, the user browses the contextual graph for tasks to be performed to reach a given goal. The user can enable the CxG reader to start the running of a practice. Then, the system enters the analysis and decision phase where different options and choices are presented to the user, who may be interested in knowing the most selected option in a similar context and situation. Once the user validates a given context, they make their decision about which branch of the contextual element to select and continue reading or running other CxG objects. The process of reading, analyzing and validating context will continue until all of the steps of the given task are completed. At that moment, the complete decision-making process terminates to enter a post-decision phase, where the analysis and evaluation of the practice will be started.

When the anonymization developer opens and reads the contextual graphs procedure of data discovery created by the data analyst (i.e. activating CxG Reader), action A1 will be executed first. Then, the Reader will suggest different options, which correspond to the three values of the contextual element C2. The KAT CxG tool displays a window to enable the developer validating their choice, as shown in Table 8.5. For instance, the developer can select contextual element value C2.2 to perform action A8, and then select other options and execute the remaining actions in the path composed of the sets A(P1) = {A1, A8, A10, A11}, C(P1) = {C2.2, C9.1} to read other potential fields that may not be detected by the discovery tool as storing personal or sensitive data. When the task execution is completed for the selected path (i.e. P1), the CxG Reader generates output files, which will be the sources that can load relevant data to the experience database in order to correct the result of the personal data discovery tool. The developer can also read the CxG edited by the data analyst to correct some fields flagged as personal data, as indicated in the path composed of the sets A(P2) = {A1, A8, A12, A13, A14}, C(P2) = {C2.2, C9.2}.

CxG Reading Steps	Reading Path 1	Reading Path 2
1	A1: Check Personal Data Classified by Discovery tool	A1: Check Personal Data Classified by Discovery tool
2	C2: List of Personal Data to check: ○ C2.1: Last Name ◉ C2.2: First Name ○ C2.3: City	C2: List of Personal Data to check: ○ C2.1: Last Name ◉ C2.2: First Name ○ C2.3: City
3	A8: Check fields marked as First Name	A8: Check fields marked as First Name
4	C9: List of fields storing First Name: ◉ C9.1: TEXT_1 ○ C9.2: TEXT_2 ○ C9.3: TEXT_3	C9: List of fields storing First Name: ○ C9.1: TEXT_1 ◉ C9.2: TEXT_2 ○ C9.3: TEXT_3
5	A10: Add the field TEXT_1 to the list of fields in the Discovery	A12: Remove from the discovery result the field TEXT_2 marked First Name
6	A11: Export the value {TEXT_1}="First Name"	A13: Set the field TEXT_2 to Not Applicable
7		A14: Export the value {TEXT_2}="N/A"

Table 8.5. *Example of paths generated by the CxG Reader*

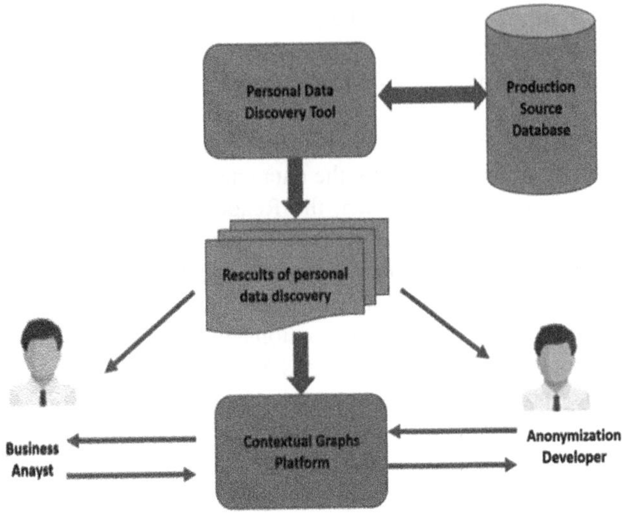

Figure 8.7. *A framework for a context-based personal data discovery tool*

In this chapter we have presented some concepts about personal data discovery in the context of anonymization. We have also illustrated how contextual graphs can be used to improve personal data discovery procedures and representing practices (i.e. contextualizing procedures). Our future work is to develop a framework for a context-based personal data discovery tool. Figure 8.7 presents the target architecture to build a framework for a context-based personal data discovery tool. The proposed tool is designed and developed to provide a uniform representation of knowledge, reasoning and contextual elements, as well as incremental knowledge acquisition and practice learning.

8.7. Conclusion

This chapter has presented how to contextualize a procedure of personal data discovery. Business analysts' practices can be simply represented by the different actions in the paths of a contextual graph. These practices correspond to checking personal data by the analyst based on the context (i.e. contents, data types, etc.) of each field in a database table. Practices are added in contextual graphs to contextualize official automated procedures for personal data discovery tools. All of the data stored in each field and classified by the automatic data discovery tool as personal will then be contextualized by analysts, using the CxG formalism representation to produce the final set of fields confirmed as storing personal data. This set of fields will then be used by developers to establish the masking rules required in the implementation (or development) phase of the anonymization process. Our work can be extended and applied to other areas, particularly in industry, medicine, computer and mobile data security.

8.8. References

Abdalla, A. (2013). Personal GIS – A tool to process our personal data, KN. *Journal of Cartography and Geographic Information*, 63(2), 86–88.

Bawar, A.S., Ahirrao, S., Churi, P.P. (2018). *Anonymization Techniques for Protecting Privacy: A Survey*. IEEE, Pune.

Bezzi, M., Montagnon, G., Salzgeber, V., Trabelsi, S. (2014). Sharing data for public security. HAL Id: hal-01061070 [Online]. Available at: https://hal.inria.fr/hal-01061070 [Accessed 7 March 2021].

Bitterman, A. (2019). Automation is your secret weapon to fast, accurate discovery of sensitive data. *Delphix* [Online]. Available at: https://www.delphix.com/blog/automated-data-discovery [Accessed 6 May 2021].

Brézillon, P. (2005). Task-realization models in contextual graphs. In *Modeling and Using Context (CONTEXT 2005)*, Dey, A., Kokinov, B., Leake, D., Turner, R. (eds). Springer Verlag, Heidelberg.

Brézillon, P. (2018). CxG-based simulation of group activity [Online]. Available at: https://www.openscience.fr/IMG/pdf/iste_muc18v2n1_5.pdf [Accessed 26 February 2022].

Brézillon, P. and Pomerol, J.-C. (1999). Contextual knowledge and proceduralized context. *Proceedings of the AAAI-99 Workshop on Modeling Context in AI Applications*, Orlando, FL.

CNIL (2021). The CNIL in a nutshell, protect personal data [Online]. Available at: https://linc.cnil.fr/sites/default/files/atoms/files/the_cnil_in_a_nutshell_2021.pdf [Accessed 1 July 2021].

Elliot, M.J., Mackey, E., O'Hara, K. (2020). *The Anonymisation Decision-Making Framework*, 2nd edition. European practitioners' guide, Manchester.

IBM (2021). Security Guardium [Online]. Available at: https://www.ibm.com/security/data-security/guardium, [Accessed 10 May 2021].

Intersoft (2021). Consulting, general data protection regulation GDPR [Online]. Available at: https://eur-lex.europa.eu/legal-content/EN/TXT/?uri=CELEX:32016R0679 [Accessed 2 May 2021].

Kudim, K.A. and Yu, G. (2019). Proskudina, methods and tools for extracting personal data from theses abstracts [Online]. Available at: https://doi.org/10.15407/pp2019.02.038 [Accessed 8 September 2021].

Microsoft (2021). Azure information protection [Online]. Available at: https://azure.microsoft.com/ [Accessed 15 April 2021].

Nassar, M., Orabi, A.A., Doha, M., Bouna, B.A. (2015). An SQL-like query tool for data anonymization and outsourcing. *International Conference on Cyber Situational Awareness, Data Analytics and Assessment (CyberSA)*, pp. 1–3.

Nightfall (n.d.). Nightfall [Online]. Available at: https://nightfall.ai/ [Accessed 10 April 2021].

Oracle (2021). Data discovery overview [Online]. Available at: https://docs.oracle.com/en-us/iaas/data-safe/doc/data-discovery-overview.html [Accessed 20 July 2021].

Prasser, F., Kohlmayer, F., Lautenschlaeger, R., Kuhn, K.A. (2014). ARX-A comprehensive tool for anonymizing biomedical data. *AMIA Symposium*, 984–993.

Pretty, T. (2020). Sensitive data classification. *Cipherpoint* [Online]. Available at: https://cipherpoint.com/blog/sensitive-data-classification/ [Accessed 26 February 2022].

Santos, I., Coutinho, S.E., Moreira, L. (2018). SMDAnonymizer: A web tool for data anonymization. *6th International Workshop on ADVANCEs in ICT Infrastructures and Services (ADVANCE'2018)*, Santiago.

Siteimprove (2020). Making a personal data inventory and data map in the GDPR era [Online]. Available at: https://siteimprove.com [Last updated: 8/3/2020].

SoftwareTestingHelp (2021). 10 best data masking tools and software [Online]. Available at: https://www.softwaretestinghelp.com/data-masking-tools/ [Accessed 10 April 2021].

Tahir, H. and Brézillon, P. (2011). Improvement of database administration by procedure contextualization. *CEUR Proceedings 4th Workshop HCP Human Centered Processes*, Genoa.

Tahir, H. and Brézillon, P. (2013). Contextual graphs platform as a basis for designing a context-based intelligent assistant system. In *CONTEXT 2013*, Brézillon, P., Blackburn, P., Dapoigny, R. (eds). Springer Verlag, Berlin, Heidelberg.

Vardalachakis, M., Kondylakis, H., Koumakis, L., Kouroubali, A. (2019). ShinyAnonymizer: A tool for anonymizing health data. *International Conference on Information and Communication Technologies for Ageing Well and e-Health (ICT4AWE)*, Prague.

Vianna, D., Yong, A., Xia, C., Marian, A.T., Nguyen, T. (2014). A tool for personal data extraction. *30th International Conference on Data Engineering Workshops*. IEEE, Chicago, IL.

Yayik, A., Aybar, V., Apik, H., İçöz, S., Bakar, B., Güngör, T. (2021). Deep learning-aided automated personal data discovery and profiling. *Turkish Journal of Electrical Engineering and Computer Sciences* [To appear in 2022].

9

Situated Management Learning

In this chapter, we present a dynamic model of situated management learning, a tool for teaching, practicing and researching management. The model is inspired by a phenomenologically and psychologically plausible philosophy of action, equally rooted in desire and belief. Its method is a comprehensive theory of individual and organizational learning that is situated in practice, community, identity and experience. It leverages three different and complementary ways of understanding, modeling and using context grounded, respectively, in practice, values and theoretical insights. It calls for situating management learning in these practical, normative and theoretical contexts, using appropriate representation formalisms. Well-chosen representation formalisms are tools for externalizing and internalizing insights and leveraging learning. We present examples of each: a contextual graph to represent practical knowledge, a values table to represent political, moral and ethical knowledge, and an insights table to represent theoretical knowledge. We show how we use the model in teaching accounting at a business school in Paris and discuss the results. Combining learning deliverables that demonstrate theoretical, practical, and political, moral and ethical insights, all linked to the management situation, encourages insightful management; management that is reasonable, practical and responsible. This research is part of an insightful management initiative.

9.1. Introduction

In this chapter, we report on a method of modeling and using context in management decision-making and problem-solving that we call a situated learning approach, in acknowledgement of our debt to Lave and Wenger's seminal work on situated learning (Lave and Wenger 1991). Management decision-making and problem-solving, as taught in business schools, involves leading (determining and

Chapter written by John HEGARTY and Régis MAUBREY.

communicating values and principles/reasons for joint action), organizing (establishing and coordinating responsibility for objectives), planning (mobilizing resources and capabilities by objective) and controlling (instituting procedures and good practices to keep on track). The insights presented in this chapter emerged from our individual experiences as practitioners and learners in science and engineering and our experience in the classroom. Whereas management is also taught in schools of public administration and technical universities, our model was developed in a business school context. This research follows our research on organizational learning (Hegarty 2013), measuring insights in learning in the classroom (Hegarty and Maubrey 2019) and the transfer of practical knowledge from one cultural context to another (Maubrey 1999).

Business managers model and use context to reason in and about business situations they face, and business organizations encourage experienced managers to share with newcomers the practical knowledge that emerges in the realization of their activity (Argote and Miron-Spektor 2011). We have previously proposed context-based intelligent assistant support as a way to leverage lessons learned in organizations (Hegarty 2013) but this approach was narrowly focused on learning as doing. In this chapter, we propose a richer view of learning, adding the dimensions of identity, community and meaning (Lave and Wenger 1991); managers do not just learn by doing, they also learn by becoming (managers), coming to belong to the (management) community and giving meaning to their experience (as a manager). This wider approach to individual and organizational learning embraces all the aspects of management knowledge: the theoretical, the practical, and the political, moral and ethical dimensions of being a manager.

We use situated learning in our classrooms to leverage students' insights into management and their own management learning. Students are encouraged to take on the role of the management decision-maker or problem-solver in business situations and share theoretical, practical, political, moral and ethical insights with their peers in small learning teams. What is important in situated learning is that each insight is linked back to the evolving management learning situation; students' understanding of the management decision-making situation is progressively enriched by insights about its history and possibilities of evolution as proposed actions are realized.

In the next section, we review the research literature relevant to management decision-making and problem-solving that underpin our situated learning approach.

9.2. Management practices, values and theoretical insights

In this section, we review the research literature relevant to the representation of management practices, insights and values. We find that there are three different types of representation, corresponding to three different types of knowledge: the practical, the theoretical, and the political, moral and ethical, each corresponding to a type of context, since context is a type of tacit collective knowledge (Brézillon and Pomerol 2001) and all personal knowledge is tacit (Polanyi 2015).

The literature relevant to the practical context of management is discussed in the following section.

9.2.1. *Management practices*

In our review of the literature on practical knowledge relevant to management, we distinguish the related topics of action, activity, practice and act.

The problem with action (in philosophy) is how to understand acting intentionally; intention has been a centrally important topic in the philosophy of action since Aristotle. We begin our discussion of planning and control with a quotation from what the philosopher Donald Davidson called "the best book on action since Aristotle":

> [...] an object is not what what is aimed at *is*; the description *under which* it is aimed at is that under which it is *called* the object. (Anscombe 2000)

Computer scientists distinguish actions and objectives; thus, planning is the search for actions that realize the objective, and understanding is the discovery of the objective of observed behavior. This common-sense approach to planning and understanding is formalized in a theory of planning and understanding (Wilensky 1983), which operationalizes Anscombe's ideas. In previous work, we illustrated an expanded version of this theory of planning, including situations, practices and lessons learned, as shown in Figure 9.1.

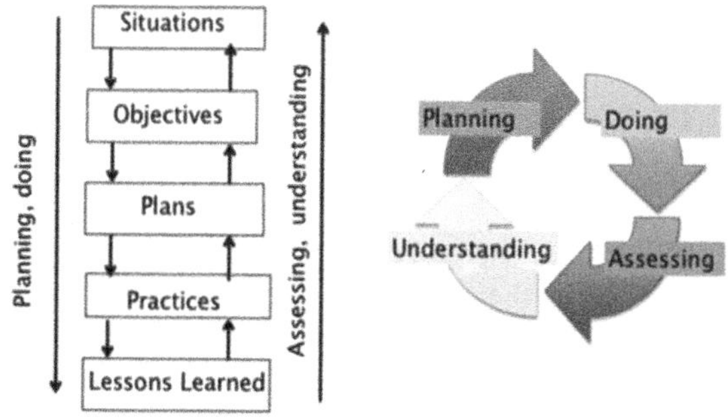

Figure 9.1. *Planning: search for actions that realize objectives (Hegarty 2013). For a color version of this figure, see www.iste.co.uk/brezillon/context.zip*

Here we enrich this notion of planning as the search for objectives, by pointing out that the discovery of actions that realize an objective is an instance of insight on the part of the planner/manager (Lonergan 1992). Thus, actions are expressions of management insight into the appropriateness of ways of using a corporation's resources and capabilities to reach its objectives. And objectives are in turn insights into the relevance of opportunities and threats for the organization.

Table 9.1 presents four types of action, each representing a different type of insight into how threats, opportunities, weaknesses and strengths are related, in a so-called TOWS chart.

TOWS	Strengths	Weaknesses
Opportunities	Actions that use resource and capability strengths to exploit an identified opportunity	Actions that reduce resource and capability weaknesses that prevent the organization from exploiting an identified opportunity
Threats	Actions that use a resource or capability strength to reduce the impact or likelihood of the identified event threatening an organizational objective	Actions that reduce resource and capability weaknesses that render the organization vulnerable to the Impact of an identified threat

Table 9.1. *Planning leverages strategic and operational insights*

Actions, whether they emerge from the planner's experience or imagination, once identified, are positioned on a timeline (sometimes called a Gantt chart); strategic and operational planning express different time horizons, strategy for longer-term actions and operations for the day-to-day running of a business. The plan is a reflective narrative of the eventual commitment of resources and capabilities by the manager decision-maker in the unfolding realization of the business opportunity.

The management control literature addresses the question of how managers ensure that activities realize the planned actions as intended. The literature from cybernetics (the science of control) is the starting point, in particular a seminal paper on behavior, purpose and teleology (Rosenblueth et al. 1943). Building on this and the work of complex open systems (Le Moigne 2005), in previous work, we used a complex systems model of the organization to situate organizational activity in its environment (Hegarty 2013). Figure 9.2 models the business organization as a system of systems: the decision-making system, the information system and the operating system (in turn a system of transforming activities that creates value). The organization is active in an evolving environment, transforming inputs from the environment into outputs of value to its customers. The organization's decision-making system uses feedback to control the realization of the transforming activities; feedback takes the form of key performance indicators (KPIs) which can be interpreted as aspectual differences in effective practices. The environment can be characterized in terms of key situational indicators (KSIs) that reflect the state of the political, economic, socio-cultural, technological, ecological and legal (PESTEL analysis) institutional environments. The KPIs and KSIs that are relevant (Wilson and Sperber 2002) for the decision-making system together form the information system, which supports communication and coordination of the activities.

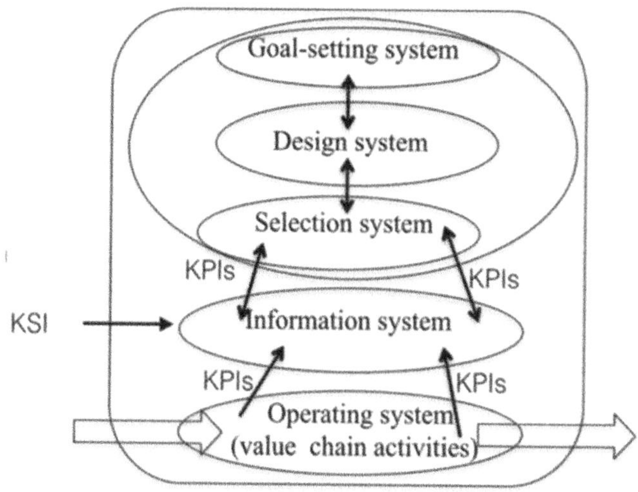

Figure 9.2. *Organization as a system of control (source: Le Moigne (2005))*

Each system is a context of organizational activity, and each activity is the context of the organizational practices that realize the activity. We use task-realization models to operationalize the decision-making system as an actor (or group agent) faced with a task to realize (Brézillon 2005). This can be done using the contextual graphs representation formalism, as shown in Figure 9.3, which recalls an application of Brézillon's theory of context to leveraging lessons learned in management practice.

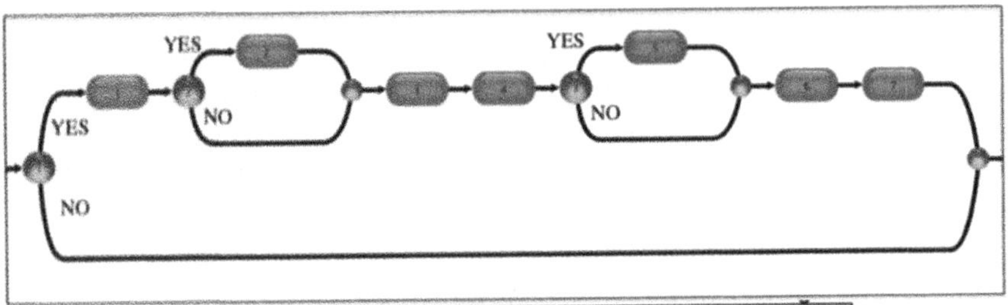

Contextual element	
CE1 (circles)	Light rail technology?
CE2	Prior work exploitable?
CE3	Complex topology?
Activity	
A 1 (ovals)	Select parameters
A 2	Exploit prior (2004) work
A 3	Generate spider's web
A 4	Sift 1 workshop
A 5	Sift 2 workshop
A 6	Consult public on route options
A 7	Appraise route options (multi-criteria framework)

Figure 9.3. *A contextual graph representation formalism (Hegarty et al. 2012). For a color version of this figure, see www.iste.co.uk/brezillon/context.zip*

The case shown in Figure 9.3 is from our study of transport planning in Dublin (Hegarty et al. 2012). The contextual graphs representation formalism was found to be useful in quite different ways at the strategic and operational levels of management. At the strategic level, the graphs were used to transpose lessons-learned board memos into a representation more amenable to post-realization analysis. At the operating level, the representation formalism is used to share practices between peers as the realization of the project unfolds.

Brézillon relates context to the evolving focus of attention as a task is realized (Brézillon 2005); the contextual graphs representation formalism can also be used to explain why actions are taken (Brézillon et al. 2002); it is a powerful experience-sharing tool in organizations (Hegarty et al. 2013).

Whereas the practices represented in a contextual graph are effective ways of doing, an instance of the practice is an act for which the actor is responsible. This brings us to the political, moral and ethical contexts of management and the representation of the associated tacit knowledge.

The literature relevant to the political, moral and ethical context of management is discussed in the following section.

9.2.2. *Management values*

In this section, we start with the contractualist theory of morality, in the narrow sense of "what we owe to each other":

> Thinking about right and wrong is, at the most basic level, thinking about what could be justified to others on grounds they, if appropriately motivated, could not reasonably reject. (Scanlon 2000)

This sense of right and wrong which is broader than justice, which has to do particularly with social institutions, and broader than obligation, which has to do with the requirements arising from specific actions and undertakings, but narrower than the broad sense that most people understand is well-suited to the business world, when combined with the notion of group agency of the corporation (List and Pettit 2011).

The corporation has obligations to its stakeholders (Freeman's wider view of the corporation's responsibility; Freeman 2001) and/or shareholders (Friedman's narrower view of the corporation's responsibility; Smith 2003).

By considering what it is to be a shareholder (investor providing a corporation with funds in return for a reasonable return) and what it is to speak in the name of a corporation (the purpose of the corporation is stated in its statutes at the moment of its registration and can be changed under certain circumstances by shareholders), we can arrive at reasons for providing a reasonable return to the shareholders that is not rejected by other stakeholders, for example the workers. Determining what is a reasonable return for the shareholders calls for judgment on the part of the corporate officer who will recommend to the shareholders the dividend to be paid out of earnings. This decision cannot (in good conscience) be reduced to the application of a rule such as "maximize shareholders' value".

In ongoing work, we are developing what we call an ethics or values table to encourage the emergence of understanding about the principles implicit in values shared in a corporation (Hegarty and Maubrey 2021). Some elements of an ethics table are presented in Table 9.2, showing shareholders and stakeholders as values, different types of fairness as principles, various functional policies/procedures as guidelines, and finally an example of an act with political, moral and/or ethical implications.

Values (Concerns)	Principles (Reasons)	Guidelines (Procedures and practices)	Acts (of interest)
Shareholders	Return on investment	Good finance practices	Decision to outsource production
Customers	Value for money	Customer relationship management	
Suppliers	Fair trade	Supplier policy	
Workers	Security of employment	Human resource policy	
Government	Contribution to society	Corporate tax policy	

Table 9.2. *Ethics table (some elements by way of illustration)*

The example in Table 9.2 comes from a course on purchasing and supply chain management that we teach at undergraduate and postgraduate levels. We encourage students to discover and share ethical, moral and political insights about the management situation using a values table; the question of outsourcing is a typically morally charged issue because of its obvious differential impact on the different types of stakeholders the corporation values. Using Scanlon's language, what the corporation owes to its shareholders, say a fair (risk-adjusted) return on investment, may be improved by outsourcing, while the workers, to whom the corporation owes, say, security of employment, are let go. It is up to the student members of a learning team to come up with and agree among themselves the principles that link the avowed values of the corporation and the actual behavior.

Values in Table 9.2 correspond to the "What?" question, principles to "Why?" and guidelines to "How?"; together they represent the "saying" of the corporation, and when not coherent with the corporation's acts or "doing", they give rise to cognitive dissonance or bad conscience of the manager. Conscience can be seen as an emotional–psychological feedback system insofar as feeling bad about acts undertaken invites us to rethink values, principles and guidelines. One way of doing this is presented in our values table, used in teaching business ethics in a way that encourages students to develop insights into what is reasonable and to take moral responsibility for their acts.

The literature relevant to the theoretical context of management is discussed in the following section.

9.2.3. *Management insights*

What managers know *tout court* (*savoir tout court*) as opposed to what they know how to do (*savoir-faire*) and how to be (*savoir-être*), we call management insight, in acknowledgment of our debt to Lonergan's seminal work on insight (Lonergan 1992). Our approach to management exploits cognitional structure (Lonergan 1967), with each of the precepts of insightful management corresponding to a successively higher level of cognitional structure, as shown below.

- be attentive: attentive managers detect relevant data;
- be intelligent: intelligent managers see patterns in the data;
- be reasonable: reasonable managers see reasons for action;
- be responsible: responsible managers make decisions they can justify.

To operationalize this notion of insightful management, we draw first on the notion of situation; management is taking place in a situation, which can be defined, in the widest possible sense, as a simultaneous multiplicity (Dewey 2018). This introduces the notion of time; situations have a history and evolve over time.

Second, we introduce the notion of agent into the situation; the manager's role as an agent in the situation is that of a decision-maker or problem-solver (Simon et al. 1987).

Third, we introduce the notion of group agency; the manager represents (speaks in the name of) a commercial corporation, that is, a legal person instituted by the shareholders but distinct from them (List and Pettit 2011).

Fourth, we introduce the notion of responsibility as it applies to group agents. The necessary and sufficient conditions for the responsibility of group agents are: 1) the choice that the agent makes must have normative significance for the agent; 2) the agent must have the ability to evaluate and judge the options available; and 3) the agent must have control over the execution of the choice. Commercial corporations, insofar as they fulfill these conditions, are group agents with responsibility for their acts (List and Pettit 2011).

Fifth, we introduce the notion of consciousness as both event and intentionality.

> Consciousness is always of something, and that something of which consciousness is conscious, I call reality, and the consciousness that is conscious of reality as an object, I call the subject, and the link

> between subject and object I call, following Husserl, intentionality. Together subject, intentionality, and object form an event that can be observed at a reflective distance by a consciousness that observes consciousness of the reality of which it is a part. (Voegelin 2001)

From these ideas, we construct the simple model of a management decision-making situation as the tension between an opportunity (intentionally aimed at as an object) and the decision-maker participating in the opportunity (as a subject, illuminated by the experience of participation). The management situation is narrated at a reflective distance as a plan.

The decision-maker is existentially conscious that there is something in the object of attention that calls for a response, and intentionality brings intelligence to bear as questions whose answers are expressed as information (which helps orient the decision-maker in the situation).

A management insight is then defined as the sudden or reflective understanding of what is going on, and what needs to be done, in a particular evolving situation confronting a corporate manager faced with a decision to make or a problem to solve.

This definition of management insight and situation allows management students, practitioners and researchers to share insights about management situations that come from many different approaches in management research.

Management insights are expressed using concepts themselves open to insights, as concepts with a history have a genealogy rather than a simple definition (Skinner 1969). Lonergan, in his classic *Insight* (Lonergan 1992), argues that insights precede concepts, "we don't just peel concepts off reality"; the occurrence of insight is an observable event, an event that can be triggered by well-chosen images.

In the classroom, we encourage students to discover and share insights about the management situation using a learning table, an example of which is given in Table 9.3.

The we-version or team learning table presented in Table 9.3 goes beyond the I-version or individual learning table presented elsewhere (Hegarty and Maubrey 2019). This change is grounded in a better understanding of understanding, as involving three terms, one person comes to an understanding with another person about something (Gadamer et al. 2004). On this view, it is the community that is the guardian of knowledge.

Insights in Purchasing and Supply Chain Mgt	We don't know	We know with help	We know	We understand
Role of procurement in the value chain				
Specification process				
Selection of suppliers process				
Contracting process				
Ordering process				
Expediting orden process				
Evaluation suppliers process				
Make or buy decision analysis				
Product /Supplier Portfolio Analysis				
Purchasing policies				
Purchasing research				
Purchasing control				

Table 9.3. *Team learning table (from purchasing and supply chain management)*

Voegelin (2001) points out how languages break forth with consciousness. In our case, management language breaks forth as communities of management decision-makers and problem-solvers engage more deeply with the reality that confronts them and of which they are a part. Management language emerges with consciousness of the content (insights) that it expresses, of the different contexts in which it is used, and of the rules that coordinate its practice. In this way, the community is the guardian of the theoretical, practical, and political, moral and ethical insights discovered and shared by the practitioners, and of the history of the development of these ideas.

In this section, we have reviewed the literature on management practices, theoretical insights and values. We reviewed different representation formalisms for the three different types of contexts: the theoretical, the practical, and the political, moral and ethical. In the next section, we return to recent philosophy of action literature that suggests a way to bring management values (degree of liking, desire) and management theoretical insights (degree of confidence, belief) together in a dynamic regulative model of action that we transpose into a dynamic model of situated management learning.

9.2.4. *Toward a dynamic model of situated management learning*

Desire and belief together dynamically regulate "acting intentionally" (Railton 2018). But our ways of doing, our practice, also reflects our communities' understanding of the right way to respond to situations (Gadamer et al. 2004). Thus, the practical knowledge communicated in the tradition of those participating in the management community is brought to each new situation by the manager. Management practice is enlightened and guided by this common knowledge, and each concrete unfolding realization is expressed in a narrative using management symbols and management concepts.

In Table 9.4, we situate the learning and practice of management in three different places: in the classroom, the boardroom and the town hall. Each actor type (planner, manager, learner) experiences a tension unfolding between an opportunity and resources and capabilities. And these are narrated in different types of plans. Table 9.4 explains how context expresses the existential consciousness that guides the actor participating intentionally in realizing an opportunity, narrated reflectively as a plan. Context is neither the opportunity nor the plan; it is whatever bears on the tension between the two, which we call a situation.

Place	Actor (participant)	Context (existential consciousness)	Content (intentional consciousness)	Process (reflective consciousness)
Town Hall	Planner	Ecosystem/ Territory's Resources and Capabilities	Development Opportunity	Development Plan
Boardroom	Manager	Business Environment/ Organization's Resources and Capabilities	Business Opportunity	Business Plan
Classroom	Learner	Management (Discipline)/ Learner's Resources and Capabilities	Learning Opportunity	Learning Plan

Table 9.4. *Situated management learning*

The next section presents an application in a management classroom.

9.3. Situated learning – an application in an accounting classroom

The method we use in the classroom is inspired by Lave and Wenger's situated learning. We ask students to form learning teams with four members, one of whom is to be the leader. We give the students some time to choose a team leader and each team chooses a business to which the learning applies. In this section, we present the case of teaching accounting to a second-year class of 24 English-speaking international students in a business school in Paris.

9.3.1. *The rules of the learning game*

We kick off the class by introducing the students to the five related preliminary concepts: business, management, corporation, information and decision-making. These concepts together form a context in the sense of Lonergan; a context is all of the relevant questions, in this case "corporate business management decision-making". We use these concepts to trigger insights and explain how to measure insight drawing a learning table, as shown in Table 9.1.

The protocol for the students is as follows.

– The theoretical deliverable for the project is summed up in the learning table with the list of insights/learning objectives of the course (see also the syllabus).

– For each insight, we should have an image that captures the insight, an application of the insight, and a consideration of the political, moral and ethical implications of using the principle. For example:

- **Image:** an image that captures the idea of value is the accounting equation, in the form owners' equity = assets minus liabilities or alternatively in the form owners' equity = capital stock plus retained earnings.

- **Application:** this shows how to use the insight, for example demonstrating that we can get the value of the business from the balance sheet that we have prepared.

- **Implication:** this discusses the political, moral or ethical implications of using that idea of value, for example is it right to consider the value of the business only in terms of shareholder value?

An example of an insight about the notion of business is the different but complementary ways of seeing a business as a set of activities and as a network of relations; information about the former will include measures of the financial performance of the activity, whereas information about the latter will involve assessing the quality of the relations between the corporation and its stakeholders. This insight is about the content of the notion, what it means, what it refers to and what it denotes.

A second example of an insight is the realization that financial accounting is about the collection, processing and use of information by investment decision-makers about a business organized in the form of a corporation; this insight concerns the context.

To encourage further interdisciplinary insights, we encourage the student teams to project themselves into a real business situation, an existing small business corporation or a plausible start-up.

Since insights can be shared, normally everyone on the team should have the same level of understanding, given sufficient time (Bloom 1974).

9.3.2. *The accounting decision-making situation*

An adequate characterization of the management decision-making situation is essential to modeling and using context in management. The definition of situation as a simultaneous multiplicity (Dewey 2018) needs to be specified as to what are the elements of the situation, the location of the elements in space and the evolution of the relations between the elements over time. Lemoigne's systems-of-systems approach is useful in introducing the notion of complexity to students who may be familiar with engineering. What works best in undergraduate management is the tension between opportunity and the entrepreneur/manager model that we specify for management decision-making.

We successively introduce the notions of business, management, corporation, information and decision-making. We characterize any business situation as an unfolding event, viewed at a reflective distance (objectively), by a decision-maker (practitioner or observer) in which the decision-maker is the subject and the business managed in the form of a commercial corporation forms the object of the decision-maker's attention. But at the same time, the students are asked to situate themselves in the role of the decision-maker intending the object of their attention. What questions must they ask and what information is required to answer these questions?

The role of the decision-maker raises the interesting distinction between the roles of the investor-shareholder and the manager-officer. The accounting information needed to make investment decisions is called financial accounting, whereas the accounting information needed to make management decisions is called management accounting. Casting the situation in this way facilitates the acquisition of insights and sharing in the team.

9.3.3. *Learning teams*

In this section, we show how this method of teaching, which is grounded in the notion of community of practice, addresses all four levels of Lonergan's theory of cognition (Lonergan 1967, 1992). The financial accounting community shares practices on what data to attend to, on what insights constitute financial accounting intelligence, on what passes for rational attitudes (thoughts and actions) and what constitutes good judgment. Students learn these things through legitimate peripheral participation in the community of accounting practitioners of which they are becoming a member by playing the role of accountant in the learning team, as well as by joining practitioner associations, attending professional events and asking teachers who are current or past practitioners to share their experience of what it is to be an accountant.

We agree with Lave and Wenger (1991) and Gadamer et al. (2004) that the community is the guardian of the norms and values of the practice of an activity. We therefore encourage the student teams to discuss the values that underpin their business plans. When students are not familiar with our ethics or values table (see Figure 9.4), we present it to them as a tool for discovering and sharing political, moral and ethical insights. Using the ethics table hones students' ability to assess the ethical dilemmas that managing a business corporation may throw up.

9.3.4. *Deliverables by the learning teams*

Each learning team is to present for their project a selection of situations encountered, each corresponding to an economic event captured in an accounting transaction. For each transaction, the learning team must analyze in an ethics table the political, moral or ethical implications of the corresponding corporate act, an image capturing each theoretical insight and a practical application of the insight in the form of a calculation or financial statement, demonstrating the practical insight and an implication.

9.3.4.1. *Deliverables that demonstrate the theoretical insight*

Students are required to present a learning table, showing their level of understanding of the theoretical insights of accounting (corresponding to the learning objectives on the syllabus of the course). The degree of chunking of knowledge is a matter for discussion by the team, but they should show at least one insight per session of the course.

9.3.4.2. *Deliverables that demonstrate the practical insight*

Students are expected to present one practical application for each insight listed in the learning table. This can be in the form of a financial statement, a journal entry for different accounting transaction types, etc.

9.3.4.3. *Deliverables that demonstrate the political, moral and ethical insights*

For each of the insights listed on the learning table, students are expected to select an accounting transaction (recorded in the journal) that corresponds to a politically, morally and ethically challenging situation, and show on a values table the values, principles, guidelines and acts analyzed.

9.3.5. *Feedback to the learning teams*

Feedback is organized in three waves: first, the members of the learning team are asked to give themselves feedback, then the class is asked for feedback, and finally the teacher completes the feedback process. Feedback emphasizes the unity of the theoretical image, the practical application, and the political, moral and ethical implications in each business situation. Students are encouraged to reflect on their learning as practice, as becoming an expert practitioner, as belonging to a community and as experience.

9.4. Results

The results of the research program that we are reporting in this chapter is a series of insights into management practice that bear on the teaching and learning of management and a series of insights about management learning and teaching that bear on the practice of management. These insights are presented in this section.

Situating management learning encourages insightful management. Reflective distance supports theoretical, political, moral and ethical learning. Representing team learning supports the sharing of management insights. Explicitly measuring shared management insight in teams encourages more sharing. Team ethics tables encourage cross-disciplinary reflective distance. Learning rules emerge from the effective practice of management learning and can be shared in learning teams to leverage learning.

Well-chosen images, adapted to the learner, make all the difference in the classroom. For example, our experience has been that the systems model and the language used in Figure 9.2 are less accessible to undergraduate students, who more easily identify with the role of a decision-maker or problem-solver in the situation model of Figure 9.4. For postgraduate students and practitioners, the systems model remains interesting, especially when the operating system is in turn modeled as a system of functionally specialized activities. For the latter, a simple academic model, such as Porter's value chain (Simatupang et al. 2017), is useful with less experienced postgraduates, and an industry benchmarking model, such as the APQC process classification framework (Cragg et al. 2007), works better with students who have more experience in business.

Insights tables and values tables are effective as team learning tools. Insights tables help the students in a learning team to structure their joint learning program, and we have found that aligning insights tables and syllabus learning objectives reassures students. Values tables encourage students to integrate thinking from other disciplines, especially politics, corporate social responsibility and business ethics. We have found that patience is critical both in the classroom as a whole and in the learning teams; everybody in the class reaches the level because everybody on each team reaches the level, if given time (Bloom 1974).

Organizing deliverables by business situation, with an image capturing theoretical knowledge, an application demonstrating practical competence and an implication demonstrating critical thinking, focuses students' attention on the relatedness of what they are learning. Students who understand that learning is more than just doing are attentive to the value of becoming an expert (practitioners), the value of belonging to a professional community and the value of their experience as legitimate peripheral practitioners.

In the next section, we will discuss the contribution of the research presented, its relation to other research and the outlook for pursuing the research program.

9.5. Discussion, outlook and related research

The results presented in the previous section encourage us to propose a more reflective approach to management, which we call insightful management, which puts insights at the heart of management, insights about management knowing, about management learning and about being a manager. We believe that information systems developers and the wider decision support research community will find the conceptual clarity of the tools presented in this chapter useful in the development of applications for discovering and sharing insights in the classroom, the boardroom and the town hall.

We are currently working on extending the complex systems model of activity to cover the requirements of institutional theorists, for example using the sustainable social–ecological systems model (Ostrom 2007). Situated sustainability assessment learning will build on our previous work in this area (Hegarty and Maubrey 2020) and is part of our insightful management initiative.

There are similarities in our approach to the approach of (cultural–historical) activity theory, the object of which is to understand the unity of consciousness and activity. In activity theory, a particular activity is a goal-directed or purposeful interaction of a subject with an object using tools. These tools are exteriorized forms of mental processes manifested in constructs, whether physical or psychological. Activity theory recognizes the internalization and externalization of cognitive processes involved in the use of tools (Nardi 1996). We expect interest from the activity theory research community in the tools that we use in the classroom to help students discover and share normative, practical and theoretical insights.

The approach to modeling and using context in management that we propose naturally progresses from the classroom to the boardroom, and to the town hall, as students of insightful management become reasonable, practical and responsible managers in business and public affairs. What we have to say about management decision-making applies equally to the decision-making of the entrepreneur and the innovator; the innovator, who has an idea for a product or service, the entrepreneur, who transforms the idea into a business, and the manager, who runs the business, play different roles in the realization of a business project. We encourage the innovation and entrepreneurship research communities to adapt our *dynamic model of situated management learning* to their domains in the form of *situated innovation learning* and *situated entrepreneurship learning*.

Insofar as these claims express a vision for the future, this chapter is also a manifesto for insightful management, entrepreneurship and innovation. In this spirit, we are currently formulating an insightful management initiative at one of the schools in which we teach.

9.6. Acknowledgments

The philosophical positions presented in this chapter are the authors' sole responsibility, but they have matured over two years of discussion in regular Zoom meetings with two Lonergan scholars, Br. Dunstan Robidoux OSB and Roland Krismer, whose erudition in philosophical and theological matters has led us to new insights in management.

9.7. References

Anscombe, G.E.M. (2000). *Intention*. Harvard University Press, Cambridge, MA.

Argote, L. and Miron, E. (2011). Organizational learning: From experience to knowledge. *Organization Science*, 22, 1123–1137.

Bloom, B.S. (1974). Time and learning. *American Psychologist*, 29, 682–688.

Brézillon, P. (2005). Task-realization models in contextual graphs. *Proceedings of the 5th International and Interdisciplinary Conference on Modeling and Using Context*, 55–68, Paris.

Brézillon, P. and Pomerol, J.-C. (2001). Is context a kind of collective tacit knowledge [Online]. Available at: http://citeseerx.ist.psu.edu/viewdoc/download?doi=10.1.1.16.991&rep=rep1&type=pdf.

Brézillon, P., Pasquier, L., Pomerol, J.-C. (2002). Reasoning with contextual graphs. *European Journal of Operational Research*, 136, 290–298.

Cragg P, Tagliavini M, Mills A (2007). Evaluating the alignment of IT with business processes in SMEs [Online]. Available at: https://ir.canterbury.ac.nz/bitstream/handle/10092/100832/Cragg%20Tagliavini%20Mills%20ACIS%202007%20%28Evaluating%20the%20Alignment%20of%20IT%20with%20Business%20Processes%20in%20SMEs%29.pdf?sequence=2&isAllowed=y.

Dewey, J. (2018). *Logic: The Theory of Inquiry*. Read Books Ltd, Redditch.

Freeman, R.E. (2001). A stakeholder theory of the modern corporation. *Perspectives in Business Ethics Sie*, 3, 38–48.

Gadamer, H.-G., Weinsheimer, J., Marshall, D.G. (2004). *EPZ Truth and Method*. Bloomsbury Publishing, New York.

Hegarty, G.J. (2013). Leveraging lessons learned in organizations through implementing practice-based organizational learning and performance improvement: An opportunity for context-based intelligent asisstant support (CIAS). PhD Thesis, NUI, Department of Accounting, Finance and Information Systems, University College Cork.

Hegarty, J. and Maubrey, R. (2019). Measuring insight in the classroom. *Proceedings of the 11th International and Interdisciplinary Conference on Modeling and Using Context*, 112–119, Trento.

Hegarty, J. and Maubrey, R. (2020). Decision making for sustainable development. *Journal of Decision Systems*, 1–14.

Hegarty, J. and Maubrey, R. (2021). Good management [Online]. Available at: https://www.openscience.fr/IMG/pdf/iste_muc21v4n1_10.pdf.

Hegarty, J., Brézillon, P., Adam, F. (2012). Practice-based performance assessment in public transport planning. In *Modeling and Using Context (CONTEXT 2013)*, Brézillon, P., Blackburn, P., Dapoigny, R. (eds). Springer, Berlin, Heidelberg.

Hegarty, J., Brézillon, P., Adam, F. (2013). The role of context in practice-based organizational learning and performance improvement. *Proceedings of the 8th International and Interdisciplinary Conference on Modeling and Using Context*, 59–72, Annecy.

Lave, J. and Wenger, E. (1991). *Situated Learning: Legitimate Peripheral Participation*. Cambridge University Press, Cambridge.

Le Moigne, J. (1999). *La modélisation des systèmes complexes*. Dunod, Paris.

List, C. and Petit, P. (2011). *Group Agency: The Possibility, Design, and Status of Corporate Agents*. Oxford University Press, Oxford.

Lonergan, B. (1967). Cognitional structure. *Collection*, 221, 239.

Lonergan, B. (1992). *Insight: A Study of Human Understanding*. University of Toronto Press, Toronto.

Maubrey, R. (1999). Etude du rôle des interactions socio-langagières dans la transmission de savoirs entre adultes : la situation de formation trans-culturelle en agriculture : cas de stagiaires africains. Thesis, Paris 8.

Nardi, B.A. (1996). *Context and Consciousness: Activity Theory and Human–Computer Interaction*. MIT Press, Cambridge, MA.

Ostrom, E. (2007). Sustainable social-ecological systems: An impossibility? [Online]. Available at: https://papers.ssrn.com/sol3/papers.cfm?abstract_id=997834.

Polanyi, M. (2015). *Personal Knowledge: Towards a Post-Critical Philosophy*. University of Chicago Press, IL.

Railton, P. (2018). Learning and doing: Toward a unified account of rationality in belief, desire and action [Online]. Available at: http://media.philosophy.ox.ac.uk/docs/lectures/locke/TT18_lec2.pdf.

Rosenblueth, A., Wiener, N., Bigelow, J. (1943). Behavior, purpose and teleology. *Philosophy of Science*, 10, 18–24.

Scanlon, T. (2000). *What We Owe to Each Other*. Belknap Press, Cambridge, MA.

Simatupang, T.M., Piboonrungroj, P., Williams, S.J. (2017). The emergence of value chain thinking. *International Journal of Value Chain Management*, 8, 40–57.

Simon, H.A., Dantzig, G.B., Hogarth, R., Plott, C.R., Raiffa, H., Schelling, T.C., Shepsle, K.A., Thaler, R., Tversky, A., Winter, S. (1987). Decision making and problem solving. *Interfaces*, 17, 11–31.

Skinner, Q. (1969). Meaning and understanding in the history of ideas. *History and Theory*, 8, 3–53.

Smith, H.J. (2003). The shareholders vs. stakeholders debate. *MIT Sloan Management Review*, 44, 85–90.

Voegelin, E. (2001). *Order and History*. University of Missouri Press, Columbia, MO.

Wilensky, R. (1983). *Planning and Understanding: A Computational Approach to Human Reasoning*. Addison-Wesley, Boston, MA.

Wilson, D. and Sperber, D. (2002). *Relevance Theory*. Blackwell, Oxford.

List of Authors

Claire ANJOU
Université TELUQ
Quebec
Canada

Jacqueline BOURDEAU
Université TELUQ
Montreal
Canada

Patrick BRÉZILLON
LIP6
Sorbonne University
Paris
France

Lamprini CHARTOFYLAKA
CRREF (EA 4538)
Université des Antilles
Guadeloupe
France

Thomas FORISSIER
CRREF (EA 4538)
Université des Antilles
Guadeloupe
France

Avelino J. GONZALEZ
University of Central Florida
Orlando
USA

John HEGARTY
IPAG Business School
Paris
France

Andreas H. MARPAUNG
University of Central Florida
Orlando
USA

Régis MAUBREY
IPAG Business School
Paris
France

Valéry PSYCHÉ
Université TELUQ
Quebec
Canada

Alain STOCKLESS
UQAM
Montreal
Canada

Hassane TAHIR
Links Consultants
Paris
France

Roy M. TURNER
School of Computing and
Information Science
University of Maine
Orono
USA

Christian WILSON
School of Computing and
Information Science
University of Maine
Orono
USA

Index

A, B

C, D

E, F, G

H, I

J, K, L

M, N, O

P, R

S, T

U, V, W

Other titles from

ISTE

in

Information Systems, Web and Pervasive Computing

2022

KARAM Elie
General Contractor Business Model for Smart Cities: Fundamentals and Techniques

2021

BEN REBAH Hassen, BOUKTHIR Hafedh, CHÉDEBOIS Antoine
Website Design and Development with HTML5 and CSS3

EL ASSAD Safwan, BARBA Dominique
Digital Communications 1: Fundamentals and Techniques
Digital Communications 2: Directed and Practical Work

GAUDIN Thierry, MAUREL Marie-Christine, POMEROL Jean-Charles
Chance, Calculation and Life

LAURENT Sébastien-Yves
Conflicts, Crimes and Regulations in Cyberspace
(Cybersecurity Set – Volume 2)

LE DEUFF Olivier
Hyperdocumentation (Intellectual Technologies Set – Volume 9)

PÉLISSIER Maud
Cultural Commons in the Digital Ecosystem
(Intellectual Technologies Set – Volume 8)

2020

CLIQUET Gérard, with the collaboration of BARAY Jérôme
Location-Based Marketing: Geomarketing and Geolocation

DE FRÉMINVILLE Marie
Cybersecurity and Decision Makers: Data Security and Digital Trust

GEORGE Éric
Digitalization of Society and Socio-political Issues 2: Digital, Information and Research

HELALI Saida
Systems and Network Infrastructure Integration

LOISEAU Hugo, VENTRE Daniel, ADEN Hartmut
Cybersecurity in Humanities and Social Sciences: A Research Methods Approach (Cybersecurity Set – Volume 1)

SEDKAOUI Soraya, KHELFAOUI Mounia
Sharing Economy and Big Data Analytics

SCHMITT Églantine
Big Data: An Art of Decision Making
(Intellectual Technologies Set – Volume 7)

2019

ALBAN Daniel, EYNAUD Philippe, MALAURENT Julien, RICHET Jean-Loup, VITARI Claudio
Information Systems Management: Governance, Urbanization and Alignment

AUGEY Dominique, with the collaboration of ALCARAZ Marina
Digital Information Ecosystems: Smart Press

BATTON-HUBERT Mireille, DESJARDIN Eric, PINET François
Geographic Data Imperfection 1: From Theory to Applications

BRIQUET-DUHAZÉ Sophie, TURCOTTE Catherine
From Reading-Writing Research to Practice

BROCHARD Luigi, KAMATH Vinod, CORBALAN Julita, HOLLAND Scott, MITTELBACH Walter, OTT Michael
Energy-Efficient Computing and Data Centers

CHAMOUX Jean-Pierre
The Digital Era 2: Political Economy Revisited

COCHARD Gérard-Michel
Introduction to Stochastic Processes and Simulation

DUONG Véronique
SEO Management: Methods and Techniques to Achieve Success

GAUCHEREL Cédric, GOUYON Pierre-Henri, DESSALLES Jean-Louis
Information, The Hidden Side of Life

GEORGE Éric
Digitalization of Society and Socio-political Issues 1: Digital, Communication and Culture

GHLALA Riadh
Analytic SQL in SQL Server 2014/2016

JANIER Mathilde, SAINT-DIZIER Patrick
Argument Mining: Linguistic Foundations

SOURIS Marc
Epidemiology and Geography: Principles, Methods and Tools of Spatial Analysis

TOUNSI Wiem
Cyber-Vigilance and Digital Trust: Cyber Security in the Era of Cloud Computing and IoT

2018

ARDUIN Pierre-Emmanuel
Insider Threats
(Advances in Information Systems Set – Volume 10)

CARMÈS Maryse
Digital Organizations Manufacturing: Scripts, Performativity and Semiopolitics
(Intellectual Technologies Set – Volume 5)

CARRÉ Dominique, VIDAL Geneviève
Hyperconnectivity: Economical, Social and Environmental Challenges
(Computing and Connected Society Set – Volume 3)

CHAMOUX Jean-Pierre
The Digital Era 1: Big Data Stakes

DOUAY Nicolas
Urban Planning in the Digital Age
(Intellectual Technologies Set – Volume 6)

FABRE Renaud, BENSOUSSAN Alain
The Digital Factory for Knowledge: Production and Validation of Scientific Results

GAUDIN Thierry, LACROIX Dominique, MAUREL Marie-Christine, POMEROL Jean-Charles
Life Sciences, Information Sciences

GAYARD Laurent
Darknet: Geopolitics and Uses
(Computing and Connected Society Set – Volume 2)

IAFRATE Fernando
Artificial Intelligence and Big Data: The Birth of a New Intelligence
(Advances in Information Systems Set – Volume 8)

LE DEUFF Olivier
Digital Humanities: History and Development
(Intellectual Technologies Set – Volume 4)

MANDRAN Nadine
Traceable Human Experiment Design Research: Theoretical Model and Practical Guide
(Advances in Information Systems Set – Volume 9)

PIVERT Olivier
NoSQL Data Models: Trends and Challenges

ROCHET Claude
Smart Cities: Reality or Fiction

SALEH Imad, AMMI, Mehdi, SZONIECKY Samuel
Challenges of the Internet of Things: Technology, Use, Ethics
(Digital Tools and Uses Set – Volume 7)

SAUVAGNARGUES Sophie
Decision-making in Crisis Situations: Research and Innovation for Optimal Training

SEDKAOUI Soraya
Data Analytics and Big Data

SZONIECKY Samuel
Ecosystems Knowledge: Modeling and Analysis Method for Information and Communication
(Digital Tools and Uses Set – Volume 6)

2017

BOUHAÏ Nasreddine, SALEH Imad
Internet of Things: Evolutions and Innovations
(Digital Tools and Uses Set – Volume 4)

DUONG Véronique
Baidu SEO: Challenges and Intricacies of Marketing in China

LESAS Anne-Marie, MIRANDA Serge
The Art and Science of NFC Programming
(Intellectual Technologies Set – Volume 3)

LIEM André
Prospective Ergonomics
(Human-Machine Interaction Set – Volume 4)

MARSAULT Xavier
Eco-generative Design for Early Stages of Architecture
(Architecture and Computer Science Set – Volume 1)

REYES-GARCIA Everardo
The Image-Interface: Graphical Supports for Visual Information
(Digital Tools and Uses Set – Volume 3)

REYES-GARCIA Everardo, BOUHAÏ Nasreddine
Designing Interactive Hypermedia Systems
(Digital Tools and Uses Set – Volume 2)

SAÏD Karim, BAHRI KORBI Fadia
Asymmetric Alliances and Information Systems:Issues and Prospects
(Advances in Information Systems Set – Volume 7)

SZONIECKY Samuel, BOUHAÏ Nasreddine
Collective Intelligence and Digital Archives: Towards Knowledge Ecosystems
(Digital Tools and Uses Set – Volume 1)

2016

BEN CHOUIKHA Mona
Organizational Design for Knowledge Management

BERTOLO David
Interactions on Digital Tablets in the Context of 3D Geometry Learning
(Human-Machine Interaction Set – Volume 2)

BOUVARD Patricia, SUZANNE Hervé
Collective Intelligence Development in Business

EL FALLAH SEGHROUCHNI Amal, ISHIKAWA Fuyuki, HÉRAULT Laurent, TOKUDA Hideyuki
Enablers for Smart Cities

FABRE Renaud, in collaboration with MESSERSCHMIDT-MARIET Quentin, HOLVOET Margot
New Challenges for Knowledge

GAUDIELLO Ilaria, ZIBETTI Elisabetta
Learning Robotics, with Robotics, by Robotics
(Human-Machine Interaction Set – Volume 3)

HENROTIN Joseph
The Art of War in the Network Age
(Intellectual Technologies Set – Volume 1)

KITAJIMA Munéo
Memory and Action Selection in Human–Machine Interaction
(Human–Machine Interaction Set – Volume 1)

LAGRAÑA Fernando
E-mail and Behavioral Changes: Uses and Misuses of Electronic Communications

LEIGNEL Jean-Louis, UNGARO Thierry, STAAR Adrien
Digital Transformation
(Advances in Information Systems Set – Volume 6)

NOYER Jean-Max
Transformation of Collective Intelligences
(Intellectual Technologies Set – Volume 2)

VENTRE Daniel
Information Warfare – 2nd edition

VITALIS André
The Uncertain Digital Revolution
(Computing and Connected Society Set – Volume 1)

2015

ARDUIN Pierre-Emmanuel, GRUNDSTEIN Michel, ROSENTHAL-SABROUX Camille
Information and Knowledge System
(Advances in Information Systems Set – Volume 2)

BÉRANGER Jérôme
Medical Information Systems Ethics

BRONNER Gérald
Belief and Misbelief Asymmetry on the Internet

IAFRATE Fernando
From Big Data to Smart Data
(Advances in Information Systems Set – Volume 1)

KRICHEN Saoussen, BEN JOUIDA Sihem
Supply Chain Management and its Applications in Computer Science

NEGRE Elsa
Information and Recommender Systems
(Advances in Information Systems Set – Volume 4)

POMEROL Jean-Charles, EPELBOIN Yves, THOURY Claire
MOOCs

SALLES Maryse
Decision-Making and the Information System
(Advances in Information Systems Set – Volume 3)

SAMARA Tarek
ERP and Information Systems: Integration or Disintegration
(Advances in Information Systems Set – Volume 5)

2014

DINET Jérôme
Information Retrieval in Digital Environments

HÉNO Raphaële, CHANDELIER Laure
3D Modeling of Buildings: Outstanding Sites

KEMBELLEC Gérald, CHARTRON Ghislaine, SALEH Imad
Recommender Systems

MATHIAN Hélène, SANDERS Lena
Spatio-temporal Approaches: Geographic Objects and Change Process

PLANTIN Jean-Christophe
Participatory Mapping

VENTRE Daniel
Chinese Cybersecurity and Defense

2013

BERNIK Igor
Cybercrime and Cyberwarfare

CAPET Philippe, DELAVALLADE Thomas
Information Evaluation

LEBRATY Jean-Fabrice, LOBRE-LEBRATY Katia
Crowdsourcing: One Step Beyond

SALLABERRY Christian
Geographical Information Retrieval in Textual Corpora

2012

BUCHER Bénédicte, LE BER Florence
Innovative Software Development in GIS

GAUSSIER Eric, YVON François
Textual Information Access

STOCKINGER Peter
Audiovisual Archives: Digital Text and Discourse Analysis

VENTRE Daniel
Cyber Conflict

2011

BANOS Arnaud, THÉVENIN Thomas
Geographical Information and Urban Transport Systems

DAUPHINÉ André
Fractal Geography

LEMBERGER Pirmin, MOREL Mederic
Managing Complexity of Information Systems

STOCKINGER Peter
Introduction to Audiovisual Archives

STOCKINGER Peter
Digital Audiovisual Archives

VENTRE Daniel
Cyberwar and Information Warfare

2010

BONNET Pierre
Enterprise Data Governance

BRUNET Roger
Sustainable Geography

CARREGA Pierre
Geographical Information and Climatology

CAUVIN Colette, ESCOBAR Francisco, SERRADJ Aziz
Thematic Cartography – 3-volume series
Thematic Cartography and Transformations – Volume 1
Cartography and the Impact of the Quantitative Revolution – Volume 2
New Approaches in Thematic Cartography – Volume 3

LANGLOIS Patrice
Simulation of Complex Systems in GIS

MATHIS Philippe
Graphs and Networks – 2nd edition

THERIAULT Marius, DES ROSIERS François
Modeling Urban Dynamics

2009

BONNET Pierre, DETAVERNIER Jean-Michel, VAUQUIER Dominique
Sustainable IT Architecture: the Progressive Way of Overhauling Information Systems with SOA

PAPY Fabrice
Information Science

RIVARD François, ABOU HARB Georges, MERET Philippe
The Transverse Information System

ROCHE Stéphane, CARON Claude
Organizational Facets of GIS

2008

BRUGNOT Gérard
Spatial Management of Risks

FINKE Gerd
Operations Research and Networks

GUERMOND Yves
Modeling Process in Geography

KANEVSKI Michael
Advanced Mapping of Environmental Data

MANOUVRIER Bernard, LAURENT Ménard
Application Integration: EAI, B2B, BPM and SOA

PAPY Fabrice
Digital Libraries

2007

DOBESCH Hartwig, DUMOLARD Pierre, DYRAS Izabela
Spatial Interpolation for Climate Data

SANDERS Lena
Models in Spatial Analysis

2006

CLIQUET Gérard
Geomarketing

CORNIOU Jean-Pierre
Looking Back and Going Forward in IT

DEVILLERS Rodolphe, JEANSOULIN Robert
Fundamentals of Spatial Data Quality

Printed and bound by CPI Group (UK) Ltd, Croydon, CR0 4YY

16/09/2026

14963134-0001